# INNOVATIVE SITE REMEDIATION TECHNOLOGY

# SOLVENT/CHEMICAL EXTRACTION

**One of an Eight-Volume Series**

Edited by
William C. Anderson, P.E., DEE
*Executive Director, American Academy of Environmental Engineers*

**1995**

Prepared by WASTECH®, a multiorganization cooperative project managed by the American Academy of Environmental Engineers® with grant assistance from the U.S. Environmental Protection Agency, the U.S. Department of Defense, and the U.S. Department of Energy.

The following organizations participated in the preparation and review of this volume:

**Air & Waste Management Association**
P.O. Box 2861
Pittsburgh, PA 15230

**American Society of Civil Engineers**
345 East 47th Street
New York, NY 10017

**American Academy of Environmental Engineers®**
130 Holiday Court, Suite 100
Annapolis, MD 21401

**Hazardous Waste Action Coalition**
1015 15th Street, N.W., Suite 802
Washington, D.C. 20005

**Water Environment Federation**
601 Wythe Street
Alexandria, VA 22314

**Library of Congress Cataloging in Publication Data**

Innovative site remediation technology/ edited by William C. Anderson
160 p. 15.24 x 22.86cm.
Includes bibliographic references.
Contents: -- [2] Chemical treatment -- [3] Soil washing/soil flushing -- [4] Stabilization/solidification -- [5] Solvent/chemical extraction -- [6] Thermal desorption -- [7] Thermal destruction -- [8] Vacuum vapor extraction
1. Soil remediation. I. Anderson, William, C., 1943- .
II. American Academy of Environmental Engineers.
TD878.I55 1994 628.5'5 93-20786
ISBN 1-883767-02-4 (v. 2)
ISBN 1-883767-03-2 (v. 3)
ISBN 1-883767-04-0 (v. 4)
ISBN 1-883767-05-9 (v. 5)
ISBN 1-883767-06-7 (v. 6)
ISBN 1-883767-07-5 (v. 7)
ISBN 1-883767-08-3 (v. 8)



Book design by Lori Imhoff

Printed in the United States of America

# CONTRIBUTORS

This monograph was prepared under the supervision of the WASTECH® Steering Committee. The manuscript for the monograph was written by a task group of experts in solvent/chemical extraction and was, in turn, subjected to two peer reviews. One review was conducted under the auspices of the Steering Committee and the second by professional and technical organizations having substantial interest in the subject.

## PRINCIPAL AUTHORS


**James R. Donnelly, QEP** Task Group Chair
Director of Environmental Services & Technologies
Davy International

**Robert Ahlert, Ph.D., P.E., DEE**
Distinguished Professor
Rutgers University

**Richard J. Ayen, Ph.D.**
Vice President & General Manager
Clemson Technical Center
Rust Federal Services Inc.

**Sharon R. Just, P.E.**
Environmental Engineer
Engineering-Science, Inc.

**Mark C. Meckes**
Physical Scientist
Risk Reduction Engineering Laboratory
U.S. Environmental Protection Agency


## REVIEWERS

The panel that reviewed the monograph under the auspices of the Project Steering Committee was composed of:

**Richard A. Conway, P.E., DEE,** Chair
Union Carbide Corporation

**Wayne M. Kachel, Ph.D., P.E.**
Martin Marietta Corporation

**Peter B. Lederman, Ph.D., P.E., DEE, P.P.**
Center for Environmental Engineering & Science
New Jersey Institute of Technology

**Robert B. Pojasek Ph.D.**
Corporate Vice President, Environmental Programs
GEI Consultants, Inc.

**Kamalesh Sirkar, Ph.D.**
Professor of Chemical Engineering
Sponsored Chair
New Jersey Institute of Technology

# STEERING COMMITTEE

**Frederick G. Pohland, Ph.D., P.E., DEE**
*Chair*
Weidlein Professor of Environmental Engineering
University of Pittsburgh

**Domy Adriano, Ph.D.**
Professor and Head
Biogeochemical Ecology Division
The University of Georgia
*Representing*, Soil Science Society of America

**William C. Anderson, P.E., DEE**
*Project Manager*
Executive Director
American Academy of Environmental Engineers

**Colonel Frederick Boecher**
Director of Risk Management
Chemical and Biological Defense Command
U.S. Army
*Representing*, American Society of Civil Engineers

**Paul L. Busch, Ph.D., P.E., DEE**
President and CEO
Malcolm Pirnie, Inc.
*Representing*, American Academy of Environmental Engineers

**Richard A. Conway, P.E., DEE**
Senior Corporate Fellow
Union Carbide Corporation
Chair, Environmental Engineering Committee
EPA Science Advisory Board

**George Coyle**
Division of Technical Innovation
Office of Technical Integration
Environmental Education Development
U.S. Department of Energy

**Timothy B. Holbrook, P.E.**
Engineering Manager
Groundwater Technology, Inc.
*Representing*, Air & Waste Management Association

**Walter W. Kovalick, Jr., Ph.D.**
Director, Technology Innovation Office
Office of Solid Waste and Emergency Response
U.S. Environmental Protection Agency

**Joseph F. Lagnese, Jr., P.E., DEE**
Private Consultant
*Representing,* Water Environment Federation

**Peter B. Lederman, Ph.D., P.E., DEE, P.P.**
Center for Env. Engineering & Science
New Jersey Institute of Technology
*Representing,* American Institute of Chemical Engineers

**Raymond C. Loehr, Ph.D., P.E., DEE**
H.M. Alharthy Centennial Chair and Professor
Civil Engineering Department
University of Texas

**Timothy Oppelt**
Director, Risk Reduction Engineering Laboratory
U.S. Environmental Protection Agency

**David Patterson**
Senior Technical Analyst
Waste Policy Institute
*Representing*, U.S. Department of Defense

**George Pierce, Ph.D.**
Editor-in-Chief
Journal of Microbiology
Manager, Bioremediation Technology Development
Cytec Industries
*Representing*, Society of Industrial Microbiology

**Peter W. Tunnicliffe, P.E., DEE**
Senior Vice President
Camp Dresser & McKee, Incorporated
*Representing,* Hazardous Waste Action Coalition

**Charles O. Velzy, P.E., DEE**
Private Consultant
*Representing,* American Society of Mechanical Engineers

**Walter J. Weber, Jr., Ph.D., P.E., DEE**
Earnest Boyce Distinguished Professor
University of Michigan
*Representing*, Hazardous Waste Research Center

# REVIEWING ORGANIZATIONS

The following organizations contributed to the monograph's review and acceptance by the professional community. The review process employed by each organization is described in its acceptance statement. Individual reviewers are, or are not, listed according to the instructions of each organization.

## Air & Waste Management Association

The Air & Waste Management Association is a nonprofit technical and educational organization with more than 14,000 members in more than fifty countries. Founded in 1907, the Association provides a neutral forum where all viewpoints of an environmental management issue (technical, scientific, economic, social, political, and public health) receive equal consideration.

This worldwide network represents many disciplines: physical and social sciences, health and medicine, engineering, law, and management. The Association serves its membership by promoting environmental responsibility and providing technical and managerial leadership in the fields of air and waste management. Dedication to these objectives enables the Association to work towards its goal: a cleaner environment.

Qualified reviewers were recruited from the Waste Group of the Technical Council. It was determined that the monograph is technically sound and publication is endorsed.

The lead reviewer was:

**Paul Lear**
OH Materials, Inc.

## American Society of Civil Engineers

Qualified reviewers were recruited from the Environmental Engineering Division of ASCE and formed a Subcommittee on WASTECH®. The members of the Subcommittee have reviewed the monograph and have determined that it is acceptable for publication.

The reviewers were:

**Cecil Lue-Hing, Sc.D., P.E.**
Metropolitan Water Reclamation District of Greater Chicago

**David T. Lordi, Ph.D.**
Metropolitan Water Reclamation District of Greater Chicago

**David R. Zenz, Ph.D., P.E.**
Metropolitan Water Reclamation District of Greater Chicago

## Hazardous Waste Action Coalition

The Hazardous Waste Action Coalition (HWAC) is an association dedicated to promoting an understanding of the state of the hazardous waste practice and related business issues. Our member firms are engineering and science firms that employ nearly 75,000 of this country's engineers, scientists, geologists, hydrogeologists, toxicologists, chemists, biologists, and others who solve hazardous waste problems as a professional service. HWAC is pleased to endorse the monograph as technically sound.

The lead reviewer was:

**James D. Knauss, Ph.D.**
Shield Environmental
Associates, Inc.

## Water Environment Federation

The Water Environment Federation is a nonprofit educational organization composed of member and affiliated associations throughout the world. Since 1928, the Federation has represented water quality specialists including engineers, scientists, government officials, industrial and municipal treatment plant operators, chemists, students, academic and equipment manufacturers, and distributors.

A qualified reviewer was recruited from the Federation's Hazardous Wastes Committee.

It has been determined that the document is technically sound and publication is endorsed.

The reviewer was:

**James A. Kent, Ph.D., P.E.**
Chemical Engineering Consultant

# ACKNOWLEDGMENTS

The WASTECH® project was conducted under a cooperative agreement between the American Academy of Environmental Engineers® and the Office of Solid Waste and Emergency Response, U.S. Environmental Protection Agency. The substantial assistance of the staff of the Technology Innovation Office was invaluable.

Financial support was provided by the U.S. Environmental Protection Agency, Department of Defense, Department of Energy, and the American Academy of Environmental Engineers®.

This multiorganization effort involving a large number of diverse professionals and substantial effort in coordinating meetings, facilitating communications, and editing and preparing multiple drafts was made possible by a dedicated staff provided by the American Academy of Environmental Engineers® consisting of:

**Paul F. Peters**
Assistant Project Manager & Managing Editor

**Karen M. Tiemens**
Editor

**Susan C. Zarriello**
Production Manager

**J. Sammi Olmo**
Project Administrative Manager

**Yolanda Y. Moulden**
Staff Assistant

**I. Patricia Violette**
Staff Assistant

# TABLE OF CONTENTS

# LIST OF TABLES

# LIST OF FIGURES

# 1
# INTRODUCTION

This monograph on solvent/chemical extraction (SCE) is one of a series of eight on innovative site and waste remediation technologies that is the culmination of a multiorganization effort involving more than 100 experts over a two-year period. It provides the experienced, practicing professional guidance on the application of innovative processes considered ready for full-scale application. Other monographs in this series address bioremediation, chemical treatment, soil washing/soil flushing, stabilization/solidification, thermal desorption, thermal destruction, and vacuum vapor extraction.

## 1.1 Solvent/Chemical Extraction

Solvent/chemical extraction is an ex situ separation and concentration process in which a nonaqueous liquid reagent is used to remove organic and/or inorganic contaminants from wastes, soils, sediments, sludges, or water. The process is based on well-documented chemical equilibrium separation techniques utilized in many industries such as oil extraction from soy beans, supercritical decaffeination of coffee, and separation of copper from leaching fluids.

Solvent/chemical extraction can be differentiated from soil washing in that soil washing[1] involves the use of dilute aqueous solutions of detergents or chelating agents to remove contaminants through desorption, abrasion, and/or physical separation, whereas SCE relies on the action of concentrated chemical agents.

1 . See the monograph in this series, *Innovative Site Remediation Technology: Soil Washing/Soil Flushing*—*Ed.*

Solvent/chemical extraction typically produces a treated fraction and a concentrated contaminated fraction, which requires further treatment to recover, destroy, or immobilize the contaminants. It may concentrate contaminants by a factor as high as 10,000:1, thereby significantly reducing the volume of material requiring further treatment or producing a concentrated stream for materials recovery.

The authors classified SCE technologies as follows:

- full-scale commercial or field-tested;
- bench-scale tested; and
- insufficient data available for evaluation.[2]

Full-scale commercial and field-tested processes are addressed in the monograph proper. Processes demonstrated at the bench-scale level that the authors deemed to warrant review are briefly addressed in Appendix B. Technologies where insufficient data were available are listed in Section 2.1 and 3.1 but are not discussed.

## 1.2 Development of the Monograph

### 1.2.1 Background

Acting upon its commitment to develop innovative treatment technologies for the remediation of hazardous waste sites and contaminated soils and groundwater, the U.S. Environmental Protection Agency (EPA) established the Technology Innovation Office (TIO) in the Office of Solid Waste and Emergency Response in March, 1990. The mission assigned TIO was to foster greater use of innovative technologies.

In October of that same year, TIO, in conjunction with the National Advisory Council on Environmental Policy and Technology (NACEPT), convened a workshop for representatives of consulting engineering firms, professional societies, research organizations, and state agencies involved in

2. The classifications were based upon information provided by process developers or suppliers and independent sources, such as the US Environmental Protection Agency. See Appendix A for a list of vendor contacts.

remediation. The workshop focused on defining the barriers that were impeding the application of innovative technologies in site remediation projects. One of the major impediments identified was the lack of reliable data on the performance, design parameters, and costs of innovative processes.

The need for reliable information led TIO to approach the American Academy of Environmental Engineers®. The Academy is a long-standing, multidisciplinary environmental engineering professional society with wide-ranging affiliations with the remediation and waste treatment professional communities. By June 1991, an agreement in principle (later formalized as a Cooperative Agreement) was reached. The Academy would manage a project to develop monographs describing the state of available innovative remediation technologies. Financial support would be provided by the EPA, U.S. Department of Defense (DOD), U.S. Department of Energy (DOE), and the Academy. The goal of both TIO and the Academy was to develop monographs providing reliable data that would be broadly recognized and accepted by the professional community, thereby, eliminating or, at least, minimizing this impediment to the use of innovative technologies.

The Academy's strategy for achieving the goal was founded on a multiorganization effort, WASTECH® (pronounced Waste Tech), which joined in partnership the Air and Waste Management Association, the American Institute of Chemical Engineers, the American Society of Civil Engineers, the American Society of Mechanical Engineers, the Hazardous Waste Action Coalition, the Society for Industrial Microbiology, and the Water Environment Federation, together with the Academy, EPA, DOD, and DOE. A Steering Committee composed of highly respected representatives of these organizations having expertise in remediation technology formulated the specific project objectives and process for developing the monographs (see page iv for a listing of Steering Committee members).

By the end of 1991, the Steering Committee had organized the Project. Preparation of the monograph began in earnest in January, 1992.

### 1.2.2 Process

The Steering Committee decided upon the technologies, or technological areas, to be covered by each monograph, the monographs' general scope, and the process for their development and appointed a task group composed of five or more experts to write a manuscript for each monograph. The task groups were appointed with a view to balancing the interests of the groups principally concerned with the application of innovative site and waste remediation technologies — industry, consulting engineers, research, academe, and government (see page iii for a listing of members of the Solvent/ Chemical Extraction Task Group).

The Steering Committee called upon the task groups to examine and analyze all pertinent information available, within the Project's financial and time constraints. This included, but was not limited to, the comprehensive data on remediation technologies compiled by EPA, the store of information possessed by the task groups' members, that of other experts willing to voluntarily contribute their knowledge, and information supplied by process vendors.

To develop broad, consensus-based monographs, the Steering Committee prescribed a twofold peer review of the first drafts. One review was conducted by the Steering Committee itself, employing panels consisting of two members of the Committee supplemented by at least four other experts (See *Reviewers,* page iii, for the panel that reviewed this monograph). Simultaneous with the Steering Committee's review, each of the professional and technical organizations represented in the Project reviewed those monographs addressing technologies in which it has substantial interest and competence. Aided by a Symposium sponsored by the Academy in October, 1992, persons having interest in the technologies were encouraged to participate in the organizations' review.

Comments resulting from both reviews were considered by the Task Group, appropriate adjustments were made, and a second draft published. The second draft was accepted by the Steering Committee and participating organizations. The statements of the organizations that formally reviewed this monograph are presented under *Reviewing Organizations* on page v.

## 1.3 Purpose

The purpose of this monograph is to further the use of innovative SCE site remediation and waste processing technologies, that is, technologies not commonly applied, where their use can provide better, more cost-effective performance than conventional methods. To this end, the monograph documents the current state of a number of innovative SCE processes.

## 1.4 Objectives

The monograph's principal objective is to furnish guidance for experienced, practicing professionals and users' project managers. The monograph is intended, therefore, not to be prescriptive, but supportive. It is intended to aid experienced professionals in applying their judgment in deciding whether and how to apply the technologies addressed under the particular circumstances confronted.

In addition, the monograph is intended to inform regulatory agency personnel and the public about the conditions under which the processes it addresses are potentially applicable.

## 1.5 Scope

The monograph addresses innovative SCE technologies that have been sufficiently developed so that they can be used in full-scale applications. It addresses all aspects of the technologies for which sufficient data were available to the Solvent/Chemical Extraction Task Group to describe and explain the technologies and assess their effectiveness, limitations, and potential applications. Laboratory- and pilot-scale studies were addressed, as appropriate.

The monograph's primary focus is site remediation and waste treatment. To the extent the information provided can also be applied to production waste streams, it will provide the profession and users this additional benefit. The monograph considers all waste matrices to which SCE can be reasonably applied, such as soils, sludges, filter cake, and other solid media.

Application of site remediation and waste treatment technology is site specific and involves consideration of a number of matters besides alternative technologies. Among them are the following that are addressed only to the extent essential to understand the applications and limitations of the technologies described:

- site investigations and assessments;
- planning, management, specifications, and procurement;
- regulatory requirements; and
- community acceptance of the technology.

## 1.6 Limitations

The information presented in this monograph has been prepared in accordance with generally recognized engineering principles and practices and is for general information only. This information should not be used without first securing competent advice with respect to its suitability for any general or specific application.

Readers are cautioned that the information presented is that which was generally available during the period when the monograph was prepared. Development of innovative site remediation and waste treatment technologies is ongoing. Accordingly, postpublication information may amplify, alter, or render obsolete the information about the processes addressed. Compared to many of the other classes of treatment technologies for which monographs are being produced, solvent extraction is a young technology, with fewer vendors that have carried their processes to full scale application. As a consequence, the reader will se a marked variation in the amount of information presented on different technologies. This variability in the quantity of information presented here on different technologies only reflects the extent to which information has been produced and was available to the authors; it is not intended to indicate a preference for one technology over another.

This monograph is not intended to be and should not be construed as a standard of any of the organizations associated with the WASTECH® Project; nor does reference in this publication to any specific method, product, process, or service constitute or imply an endorsement, recommendation, or warranty thereof.

# 1.7 Organization

This monograph and others in the series are organized under a uniform outline intended to facilitate cross reference among them and comparison of the technologies they address. Chapter 2.0, Process Summary, provides an overview of all material presented. Chapter 3.0, Process Identification, provides comprehensive information on the processes addressed. Each process is fully analyzed in turn. The analysis includes a description of the process (what it does and how it does it), its scientific basis, status of development, environmental effects, pre- and posttreatment requirements, health and safety considerations, design data, operational considerations, and comparative cost data to the extent available. Also addressed are process unique planning and management requirements and process variations. Case studies of three of the processes are set forth in Appendices C, D, and E.

Chapter 4.0, Potential Applications, Chapter 5.0, Process Evaluation, and Chapter 6.0, Limitations, provide a synthesis of available information and informed judgments on the processes. Each of these chapters addresses the processes in the same order as they are described in Chapter 3.0. Chapter 7.0, Technology Prognosis, identifies indications of the likely expanded applications of SCE.

# 2 PROCESS SUMMARY[1]

## 2.1 Identification of Processes

Solvent/chemical extraction (SCE) technologies, based on well-established scientific principles, employ a broad range of solvents to separate contaminants from soils, sludges, sediments, and wastewater. The following are the full-scale commercial and/or field-tested SCE processes evaluated in this monograph:

- Basic Extractive Sludge Treatment (B.E.S.T.) Process;
- CF Systems;
- Carver-Greenfield Process;
- Extraksol Process;
- Low Energy Extraction Process (LEEP®)
- NuKEM Development (NKD) Process; and
- Soil Restoration Unit (S.R.U.).

The following are the emerging processes demonstrated at the bench scale that are briefly addressed in Appendix B:

- Chemical Waste Management Adiabatic Process for the Extraction of Sludges (APES);
- Phoenix Miljø Contex Process;
- Henkel Liquid Ion Exchange (LIX) Process;

1. This chapter is a summary of Chapters 3.0 through 7.0. Sources are cited, where appropriate, in those chapters.

- Davy R & D Resin-in-Pulp/Carbon-in-Pulp (RIP/CIP) Process; and
- Martin Marietta Soilex Process.

The following technologies were identified but insufficient data were available to evaluate these technologies:

- Environment Canada's Soil Treatment Process;
- Extrapure Process by EM&C; and
- Solvent Extraction for Dredged Sludges by SRE, Inc.

## 2.2 Scientific Basis

Solvent/chemical extraction effects the preferential separation of one or more constituents from one phase into a second phase. In classical chemical engineering terms, SCE is the term applied to the transfer that occurs between two liquid phases, or between a solid and a liquid phase.

In a conventional liquid-liquid contacting system, the best separation of solute removal that can be effected is determined by the relative solubilities of the solute in the two liquid phases. The ratio of the solute concentrations in the two phases at equilibrium is the equilibrium distribution coefficient. In the very poorly or partially miscible liquid-liquid cases, contaminant transfer is a function of relative solubilities and the equilibrium distribution coefficient.

When a substrate is transferred from a solid to the liquid phase, the action is called leaching. Solvent/chemical extraction is the controlled leaching of contaminants from soils, sediments, and solid wastes through use of organic solvents or nonaqueous liquids.

Solvent/chemical extraction processes used for soil/sediment cleanup typically employ a solvent which extracts both water and organics into the liquid phase. Subsequent steps involve first separating the liquid phase from the solids, then separating the water and organic phase, and finally, separating the contaminants from the solvent. As such, the extraction of the contaminant from a solid phase involves only the equilibrium of the contaminant with the solvent.

# 2.3 Process Description

## 2.3.1 Basic Extractive Sludge Treatment Process

The B.E.S.T. Process, developed by the Resources Conservation Company (RCC), uses triethylamine (TEA) as the organic solvent. Below 18°C (64°F), TEA is miscible with water and is used in dewatering solids and partially removing organic contaminants. The remaining organic contaminants are removed by TEA heated to 55° to 80°C (130° to 176°F), a temperature range in which TEA is liquid and slightly immiscible with water resulting in two liquid phases.

The process is carried out in a washer/dryer unit consisting of a steam-jacketed vessel with a horizontal mixer shaft. Mixing time varies from 5 to 15 minutes per extraction.

RCC has constructed a skid-mounted, 44-kg/day (100 lb/day) pilot plant and a nominal 91-tonne/day (100 ton/day) full-scale, transportable system.

The full-scale system was demonstrated at one site, and achieved a maximum throughput of 64 tonne/day (72 ton/day). Throughput is expected to vary with the composition of the feed material. The need for high-removal efficiencies may necessitate the use of multiple extraction stages, further reducing rated throughput.

Solids may have to be crushed and screened to meet particle size requirements (<2.5 cm (1 in.) in diameter). While the pilot plant can treat pumpable and non-pumpable solids, the full-scale system treats only pumpable waste, such as oily sludges. A portable full-scale system yet to be constructed will be capable of handling wet or dry solids. To maintain TEA in a nonionized state, the pH of the extraction mixture must be between 10.5 and 11.

Solids discharged from the B.E.S.T. Process are dry and oil-free, although some solvent and/or organic contaminants may remain. Solids meeting cleanup criteria may be returned to the land. The water fraction may contain low concentrations of organic contaminants and/or residual solvent, but may be acceptable for discharge to a publicly-owned treatment works (POTW) without further treatment. Alternatively, when required, the wastewater may be treated by liquid phase carbon adsorption.

The organic contaminants are concentrated in the oil raffinate or fraction and are typically recovered by distillation for recycling or treatment. Depending on which contaminants are present, treatment options may include incineration, chemical dehalogenation, or recycling.

Triethylamine is a highly flammable solvent and, therefore, must be transferred with nonsparking equipment. Restricted-access zones are required around the extraction plant. During full-scale operations previously discussed, TEA leaks caused severe odor problems.

### 2.3.2 CF Systems

In the CF Systems Process, liquefied gases and supercritical fluids are used as extraction solvents to separate organic species from wastewater, sludge, and/or contaminated soil. Liquefied carbon dioxide is generally used for aqueous solutions. Propane, or a mixture of propane/butane, is often used for sediments, sludges, and soils.

In a typical operation employing liquefied propane, contaminated sediments are fed top-down into a high-pressure contactor. Compressed propane at 20°C (68°F) passes upward, counterflow to the solids, and dissolves organic matter from the feed. The clean sediment is removed from the contactor and the propane, which contains the organic contaminants, is passed to a separator. The propane is vaporized, recompressed, and recycled to the contactor. Contaminants and natural organic matter are removed from the separation vessel and disposed or reused. The same concept applies in the use of supercritical carbon dioxide or liquefied light hydrocarbon gas mixtures.

Bench-scale process units are used to size pilot- and full-scale facilities. These units employ a single-contact stage that is applied repetitively to estimate the number of contact stages necessary to attain the desired extraction efficiency. In bench-scale studies conducted with supercritical carbon dioxide, CF Systems (CFS) reported extraction efficiencies ranging from 95% to 100%.

On a pilot scale, CFS demonstrated a trailer-mounted unit for treating soils, sludges, and semisolids with liquefied propane and a propane/butane mixture. This portable unit, the Pit Cleanup Unit (PCU), utilizes two stages of countercurrent extraction with solid-liquid separation between extractors. The PCU has been used to treat sediments contaminated with polychlorinated biphenyls (PCBs), soils and sludges at nine refineries, petrochemical plants, and treatment, storage, and disposal (TSD) facilities. Data from pilot-scale studies in which a number of contaminated soils and wastes were treated with liquefied light hydrocarbon gases indicate that Best Demonstrated Available Technology (BDAT) standards were attained. Concentrations for many contaminants were below detection limits after treatment.

Propane extraction has been performed on a commercial scale using a plant having a capacity of about 23 tonne/day (25 ton/day) at a petroleum refinery for sludge treatment. Analyses indicate that the product from commercial liquefied propane extraction of refinery sludge conforms to BDAT standards for volatile and semivolatile organic compounds. Extraction efficiencies varied from 80% to nearly 100%.

The extent of pretreatment necessary prior to supercritical extraction is specific to the particular site and material to be treated. To assure a relatively homogeneous feed, solids removal, water addition, and mixing may be necessary. Particles greater than 19 mm (0.75 in.) may need to be removed and a suitable feed temperature must be established.

Extracted contaminants containing concentrated organic matter must be contained, handled, stored, and transported in accordance with the Resource Conservation and Recovery Act (RCRA) requirements. The wet sediment product must be dewatered and the water returned for feed stream preparation, thereby reducing the need for subsequent wastewater treatment.

Propane is highly combustible and poses an explosion threat. All electrical equipment must be explosion-proof; potential sources of ignition must be eliminated where explosive gases or gas mixtures serve as the supercritical extractant. For personnel handling treated and untreated wastes, Occupational Safety and Health Act (OSHA) Level B or Level C protection is recommended depending on site conditions; Level C protection is recommended, as a minimum, for extraction process operators. High pressure operations impose a greater need for leak testing, monitoring, and pressure vessel design detail than for conventional, domestic propane application. Process systems must be monitored constantly to detect pressure losses and emissions to the atmosphere. In addition, units in the recycle train must be monitored to detect and minimize losses of contaminant to the environment.

Before a product is discharged to the environment, it must be treated to eliminate hazardous characteristics. The enriched contaminant stream must be completely destroyed, be stabilized for long-term storage, or its toxicity must be substantially reduced.

### 2.3.3 Carver-Greenfield Process

The Carver-Greenfield Process is a drying and solvent extraction treatment process that separates mixtures into solids, oil, and water, while extracting organics using a carrier oil or solvent. First, debris is separated, and then the feed particles are ground to less than 6 mm (0.25 in.) and slurried with a carrier oil or solvent to extract indigenous oils and soluble organics. The water in the slurry is evaporated in two to four multi-effect evaporators. The vapors from the evaporation step are condensed, and the water and carrier oil/solvent condensate are sent to an oil/water separator. The majority of the carrier oil/solvent is separated from the feed solids by centrifuging. The solvent is removed from the centrifuge cake by heat stripping. The used carrier solvent is distilled to recover the carrier oil/solvent and to separate the indigenous oils/organics. Some designs incorporate only some of the foregoing steps.

Originally developed in the 1950s, the Carver-Greenfield Process has been used commercially to treat meat rendering wastes and nonhazardous municipal and industrial wastes. Other commercial-scale applications include treatment of petrochemical activated sludge, biosludge and wool scouring waste, dye wastes, alum sludge, digested and undigested municipal wastewater sludges, and oily industrial sludge. In recent years, tests have been conducted to determine whether the process was suitable for treating soils and sludges contaminated with organic compounds.

Over 85 commercial-scale units have been licensed worldwide. For those employed to treat chemical waste and sudges, feed solids generally range from 2% to 20%, and treatment capacity ranges from 0.3 to 382 dry tonne/day (0.33 to 420 dry ton/day). A pilot-scale unit, employed in recent treatability tests of organic contaminated soils and sludges, has a capacity of 45 kg/hr (100 lb/hr).

The process requires 480-volt, three-phase electric power. For treating a waste consisting of 52% solids, 17% indigenous oil, and 31% water (by weight), US EPA estimated that 1.7 billion Joule would be used per tonne of waste (1.47 MM Btu/ton). Nitrogen consumption (for deoiling) was estimated to be 45.9 standard $m^3$/hr (1,620 standard $ft^3$/hr).

Several residues result, each requiring further treatment or disposal. The concentrated indigenous oil and organics may be burned or refined and reused. Although water produced by the process is substantially free of solids and oils, it usually requires further treatment. Generally, clean, dry solids remaining after treatment may be used for backfill or sent to a landfill. If inorganic compounds such as metals remain, the waste may need further treatment. If the waste feed is a listed waste, the treated solids must be delisted prior to disposal.

Potential environmental impacts include the effects of air emissions, dust releases, and hazards in transporting materials. The primary environmental impacts occur during removal of recovered indigenous oil and contaminated solvent from the site. The closed configuration of the process reduces the potential for air emissions. To mitigate fugitive dust emissions, the final product may have to be treated with additives.

### 2.3.4 Extraksol Process

The Extraksol Process is a transportable system that uses proprietary solvents, individually or blended together, for batch extraction of contaminated soils. Extraction of organic contaminants is conducted in three phases: washing, drying, and solvent regeneration.

The extraction vessel is a mix tank without an internal agitator. The developer claims that absence of an internal mixing system enables the processing of large, nonporous solids, up to 0.6 m (2 ft) in diameter. Porous solids must be no larger than 5.1 cm (2 in.). Some waste streams may need to be dewatered prior to treatment. Agitation will break down porous process solids during the initial extraction with hydrophilic solvents. After each extraction, the vessel's rotation is stopped, and the solids are allowed to settle. The contaminated solvent is decanted and pumped to a still for solvent recovery. The number of wash cycles and the wash cycle time may be varied.

The solvent is drained from the solids, which are then heated by introducing hot nitrogen gas and steam into the extractor. The heated gas strips the solvent from the solids as the extraction vessel is rotated. After the gas is removed by vacuum, it is discharged to a condenser. The contaminated solvent is pumped to a distillation unit where low boiling point components are volatilized, condensed, and collected for reuse as clean solvent. The remaining liquid components flow to a settling compartment. Decant from this vessel is discharged as wastewater, which may require additional treatment. Still bottoms contain the concentrated organic fraction which requires further treatment, such as incineration or dehalogenation, depending on its chemical makeup.

Bench-scale treatability tests should be conducted to determine whether the Extraksol Process is applicable to a specific waste stream. The bench-scale tests are conducted in a 0.23-m$^3$ (2.5 ft$^3$) rotary mixer that simulates the washing cycle.

The Sanivan Group has constructed two transportable full-scale systems—one with a capacity of 0.9 tonne/hr (1 ton/hr) and another with a capacity of 5.5 to 9 tonne/hr (6.2 to 10.1 ton/hr). Although the 0.9-tonne/hr system requires less than 28 m$^2$ (300 ft$^2$), the standard restricted access zone associated with the process significantly expands the space required.

To minimize potential air emissions, process gases are vented through an activated carbon filter before being released to the atmosphere. Process wastewaters may contain low concentrations of contaminants, necessitating additional treatment prior to discharge. Chelating agents may be used with the solvent to mobilize lead. Chelated lead would be concentrated in the wastewater fraction and require additional treatment before disposal.

### 2.3.5 Low-Energy Extraction Process

In the Low-Energy Extraction Process (LEEP®), contaminants are leached from the solid matrix through use of a hydrophilic leaching solvent and then concentrated using either a hydrophobic stripping solvent or distillation. ART International, Inc. has developed two types of plant processes. The "LEEP-Tar-plant" is targeted for coal tars and related compounds, while the "LEEP-PCB-plant" is intended to treat PCBs and related compounds. The two plants have similar mechanical configurations, but differ in solvent usage and recovery.

Bench-scale studies have been conducted, and a pilot plant with a nominal throughput of 91 kg/hr, (200 lb/hr) was completed in February 1992. ART International, Inc. has conducted additional bench- and pilot-scale tests since then to quantify operational parameters, and is in the process of developing design criteria for a commercial-size LEEP-Tar-plant. The developer anticipates a plant to be commercially operational by early 1996.

The commercial-size plants are designed to be closed loop systems capable of handling particle sizes up to 203 mm (8 in.) in diameter. If feed materials contain free water (i.e., moisture content exceeding 50 to 70%), it is more economical to pretreat the materials to separate the liquid and solid matrices. For example, typical pretreatment of high moisture content materials such as sediments includes centrifugation or filtration. In the full-size plant, solids will be fed through a grizzly/vibrating screen which rejects oversized (>203 mm(8 in.)) material, and pieces of metal will be removed. The solids will then be crushed to 13 to 25 mm (0.5 to 1 in.) in size and fed to the leaching operation. The currently available pilot-scale plant is limited to treating particles less than 6 mm (0.25 in.) in size. However, larger materials can be crushed for pilot-scale testing.

Leaching is performed at atmospheric pressure in a continuous solid/liquid countercurrent contactor. The leaching solvent in the PCB-plant is acetone, and in the Tar-plant the leaching occurs with acetone and a proprietary hydrophobic solvent. The contaminants are separated from the leachates and concentrated either by liquid-liquid extraction using kerosene (PCB-plant) or by distillation (Tar-plant). The kerosene with the contaminants is collected for off-site disposal, and the refined tar from the distillation is collected for potential commercial use.

Residual solvents associated with the leached soil are removed in a continuous dryer and are internally recycled. Any water which is leached from the contaminated solids is separated from the acetone by distillation and treated by carbon adsorption. The soil and water are then recombined in the final process step and the moist soil is discharged.

### 2.3.6 NuKEM Development Process

NuKEM Development (NKD) is developing two waste treatment processes employing solvent extraction. One process, using mixer/settlers, is targeted primarily for remediation of contaminated soil. The second process uses a continuous extraction column and is targeted for treating wastewaters and sludges from petroleum refineries.

Contaminated soil from the excavation is screened and reduced in size to <5 cm (2 in.) before it is fed into a mixer, where it is combined with a proprietary solvent and a proprietary chemical reagent. The slurried soil leaving the mixer is then fed to a countercurrent extraction system in which the organic contaminants are progressively removed as they pass through the extraction stages. Three to five stages of extraction are usually adequate to meet contaminant target levels.

The treated soil is fed to a solvent dryer to recover the solvent. The contaminated solvent (extract) is withdrawn from the first mixer/settler and is processed in a distillation system in which the contaminants are separated from the solvent. Personnel exposure to the waste and refinery-derived solvent, which is flammable, should be minimized.

The soil remediation process produces clean soil and two by-product streams — one consists of debris initially removed from the soil, and the other is the concentrated organic waste. It is possible that very light organic compounds may be present and build up in the solvent stream, which is recycled. To prevent contamination of nearby aquifers, the solvent must be thoroughly stripped from the treated soil before it is returned to the environment. Solvent losses to the atmosphere should not be significant.

A variation of this process is being developed to treat refinery waste streams. No dewatering is required prior to treatment. The extraction is carried out in a multistage column extractor. Solids and water flow down the column, while solvent flows from the bottom to the top, extracting oil from the refinery waste as it rises. The oil-laden solvent exits the top of the column and is fed to a fractional distillation column for recovery of the solvent. Oil-free solids and water exit the bottom of the column and are pumped to a solvent stripper. The resulting slurry is pumped to a filter. The filter cake is stabilized, if necessary, and sent to a landfill.

Pilot-scale studies are underway for the soil decontamination and refinery waste processes. The refinery waste pilot plant had a throughput rate of 320 to 640 L/day (2 to 4 bbl/day), but was operated at higher rates from time to time. From the results of the studies, NKD has been able to make projections about key features of a plant intended to treat oily wastes of a 16 MM L/day (100,000 bbl/day) oil refinery. A refinery of this size will generate approximately 10,000 tonne/yr (11,000 ton/yr) of oily wastes.

### 2.3.7 Soil Restoration Unit

The Soil Restoration Unit (S.R.U.) uses the same three process steps as the Extraksol Process: washing, drying, and solvent regeneration. In this process, however, unlike batch processes, the solids are conveyed through a specially designed extraction system where solids are mixed with solvent(s). The S.R.U. Process consists of eight unit operations. The specific solvent(s) to be used and the time required for extraction of wastes is determined during bench-scale treatability tests. The process can be fine-tuned based on results in actual operation.

The system can accommodate solids up to 7.6 cm (3 in.) in diameter, therefore, solids may have to be screened and crushed. Waste streams with high concentrations of contaminants may be premixed with the selected solvent to reduce overall time for extraction.

As solvent mixes with the solids, organic substances are removed. The contaminated solvents are continuously flushed with fresh solvent, causing solute concentrations in the solids to equilibrate with solute concentrations in the liquid. Since the fresh solvent contains no solute, removal is limited by diffusion or desorption of the contaminants from the solid particles. Chelating agents may be used with the solvent during the wash cycles to mobilize specific inorganic contaminants, such as metals. In the solids drying unit, residual solvent is removed and a clean, dry product results.

Terra-Kleen Corporation constructed a self-contained mobile S.R.U. capable of treating up to 1.8 tonne/hr (2 ton/hr) of contaminated solids. The mobile unit is contained in two trailers and occupies approximately 93 $m^2$ (1,000 $ft^2$), not including space for support equipment. A full-scale system has been constructed to treat approximately 2,700 tonne (3,000 ton) of soils contaminated with PCBs and chlorobenzenes at a Superfund site.

High moisture content solids can reduce the effectiveness of the hydrophilic solvents and increase energy costs for the distillation step. Flammable solvents are employed; National Fire Protection Association (NFPA) standards are applicable.

## 2.4 Potential Applications

The following section presents summaries of potential applications for the various SCE processes. Potential applications are based on data supplied by the process vendors or from US EPA SITE reports.

### 2.4.1 Basic Extractive Sludge Treatment Process

Numerous B.E.S.T. Process bench-scale treatability tests have been performed on soils, sludges, and sediments contaminated with PCBs, polyaromatic hydrocarbons (PAHs), pesticides, and other semivolatile and volatile organic contaminants. Highest removal efficiencies were achieved with solids that had high initial organic contaminant concentrations, but many of the treated solids still contained a significant amount of the contaminant. The treatability tests, in almost all cases, demonstrated the process' ability to meet cleanup goals. The B.E.S.T. Process may also be used to treat sludges from petroleum refineries and petrochemical operations.

### 2.4.2 CF Systems

Liquefied carbon dioxide is generally used for aqueous solutions, such as process water and wastewater; light hydrocarbons are recommended for sludges, sediments, and soils. Supercritical technology can be applied to a large variety of organic contaminants. The developer has described a number of possible applications in solid and semisolid waste treatment, wastewater treatment, and pollution prevention.

### 2.4.3 Carver-Greenfield Process

The Carver-Greenfield Process can be used to remove oil-soluble organics from soils, sludges, and other wastes. It can also be used to dry aqueous mixtures and to treat wastes contaminated with organics, especially wastes with high water content.

In some designs, commercial-scale units incorporate parts of the Carver-Greenfield Process for unique applications. In one design, for example, refinery sludge was dewatered with the Carver-Greenfield Process prior to coking.

### 2.4.4 Extraksol Process

The Extraksol Process can be used to treat soils, sludges, and sediments contaminated with volatile organic contaminants (VOCs), semivolatile organic contaminants (SVOCs), oils, and greases. The number of wash cycles and their periods may be varied in order to remove specific contaminants from solids. Considerable variability has been observed with respect to removal efficiencies for oil and greases, as well as for specific contaminants. Therefore, treatability testing is encouraged prior to selecting this technology for site remediation.

### 2.4.5 Low-Energy Extraction Process

The LEEP® can be used to remove coal tar and PCBs and related compounds, such as creosote, petroleum hydrocarbons, PAHs, pesticides, wood-preserving chlorophenol compounds, and pesticides from contaminated soils, sludges, and sediments. The LEEP technology has been used to treat river and harbor sediments, various topsoils, clay subsoil, and foundry sand.

### 2.4.6 NuKEM Development Process

Data have been published only on the removal of PCBs from soils and sludges, but it is expected that the NKD Process will be effective in removing a wide range of volatile organics, semivolatile organics, pesticides and their intermediates, petroleum hydrocarbons, and other organics. The petroleum refinery sludge treatment version of the process has been shown to be effective in treating sludges from American Petroleum Institute (API) separator, dissolved air flotation, and slop oil sludges.

### 2.4.7 Soil Restoration Unit

A pilot-scale system was effectively used to treat PCB-contaminated sandblasting sand at one site. Another test, conducted on sandy loam contaminated with PCBs, resulted in substantial removal of the contaminants, although according to the developer, the uninsulated system suffered considerable reduction in extraction efficiencies. Soils contaminated with diesel fuel were also treated by this process.

## 2.5 *Process Evaluation*

### 2.5.1 Levels of Contaminant Removal

The seven processes studied have been shown to be effective in removing a wide range of organic contaminants from a number of different feed materials. Contaminant concentration factors as high as 10,000:1 have been measured, this implies that these processes can provide a significant reduction in the volume of contaminants requiring further treatment.

Removal efficiencies and levels of reduction vary among the processes depending on the nature of the process, number of extraction stages, type and concentration of contaminants present, and the nature of the medium to be treated. Efficiencies in removing all organic contaminants above 90% are often reported. Residual levels below 1 ppm are also reported, but are highly dependant on the matrix and contaminant type and concentrations. In addition, a number (6 to 8) of extraction stages may be required to effect those results.

### 2.5.2 Status of Development

All seven processes have undergone extensive testing at the bench- and pilot-scale level, and five processes have full-sized commercial-scale systems. Three systems (CF Systems, B.E.S.T., and Carver-Greenfield) have been demonstrated under US EPA Superfund Innovative Technology Evaluation (SITE) demonstration programs and are fully documented. Results of test programs and evaluations have been mixed. In a number of cases, the systems have met or exceeded test objectives, while in other cases they have not. Therefore, treatability testing is required for most applications so that site-specific design parameters can be determined. Based on results of treatability testing or similar applications, suppliers are offering their systems to cover a wide variety of applications.

### 2.5.3 Secondary Environmental Impacts

As a separation process, SCE does not destroy organic contaminants but produces a concentrated contaminant fraction, a treated solids fraction, and a wastewater stream. The concentrated contaminant fraction is likely to be a hazardous waste and, as such, is subject to hazardous waste regulations. Incineration or another means is employed to destroy this fraction. The treated solids fraction and wastewater may contain residues of the organic contaminant and extraction fluid. Depending on the matrix treated, and the cleanup requirements, further treatment and/or several extraction stages may be required. Treatability studies can be helpful in assessing the matrix-specific levels of residuals for the various SCE processes and the potential requirement, if any, for further treatment.

### 2.5.4 Costs

Estimates of unit costs of the systems evaluated range from \$105 to \$770/tonne (\$95 to \$700/ton) see table 5.1 on page 5.3. Costs depend on the type of contaminant being treated and its concentration and the type of feed material. Treatability studies and site-specific cost evaluations are required to prepare meaningful cost estimates. Reported costs appear to be competitive with alternative remedial technologies.

The US EPA has published detailed cost estimates for the CF Systems Process, the Carver-Greenfield Process, and the B.E.S.T. Process.

## 2.6 *Limitations*

### 2.6.1 Site/Matrix Considerations

Solvent/chemical extraction processes are not normally effective in removing inorganic contaminants, such as heavy metals. Although highly effective in removing PCBs and PAHs, they may not be effective in removing hydrophilic and high molecular weight organic compounds. High concentrations of indigenous organic compounds in the feed can reduce extraction efficiency and processing rates.

### 2.6.2 Residue Treatment

Solvent/chemical extraction produces three residue streams — treated solids fraction, concentrated contaminant fraction, and wastewater. The cleaned solids fraction may contain residual contaminants and extraction fluid, as may the wastewater stream, along with soluble heavy metals. The concentrated contaminants fraction may often require additional treatment to assure destruction of toxic organic compounds or to prepare the fraction to be recycled.

### 2.6.3 Process Risks

Organic extraction fluids present fire hazards. Some extraction fluids are toxic organic compounds, subject to hazardous waste regulations for storage, use, transportation, and disposal.

### 2.6.4 Reliability

Most data reported for SCE technologies were generated by bench-scale, pilot-scale, or demonstration plants. Little long-term data on commercial plants are available for use in evaluating reliability. More commercial applications treating run-of-site feed are needed to demonstrate the feasibility of SCE in handling variations in feed properties. Until such data become available, extensive site-specific treatability testing is warranted when applying these technologies.

## 2.7 Technology Prognosis

The separation and concentration technique of SCE has been successfully applied in a wide variety of industries, including food processing, pharmaceuticals, fine chemicals, and mining and minerals processing. Many of the benefits of SCE accruing in these industries will apply in the treatment of soils, sludges, sediments, and wastewater. As additional committed systems are brought on line, process uncertainties will be reduced and treatment costs should decrease. As additional process concepts are developed, SCE technology will be applied more widely, leading to reduced process costs.

# 3 PROCESS IDENTIFICATION AND DESCRIPTION

## 3.1 Identification of Processes

Identification of potential technologies for evaluation in this monograph was based on information gathered from a wide variety of published and unpublished sources (US EPA 1987; 1990c; 1992c, d, e) as well as through personal contacts. The quantity and quality of available information for each technology vary greatly. Some potential technologies identified have undergone extensive testing at bench-, pilot-, and demonstration-scale levels and are ready for commercial application. Others are emerging technologies requiring additional development and, for some, development has been discontinued.

A number of solvent/chemical extraction (SCE) processes identified are undergoing testing and/or have been evaluated in the United States Environmental Protection Agency's (US EPA) Superfund Innovative Technology Evaluation (SITE) program. Three technologies have been evaluated under the SITE Demonstration program and test reports published. These are:

1. Resource Conservation Corporation, Basic Extractive Sludge Treatment (B.E.S.T.) Process (US EPA 1988);
2. CF Systems, Organic Extraction Unit (US EPA 1990a); and
3. Dehydro-Tech's Carver-Greenfield Process (US EPA 1992a).

These and other processes have been shown to be effective in separating and concentrating organic contaminants such as polychlorinated biphenyls (PCBs), pentachlorophenol (PCP), pesticides, volatile organic compounds (VOCs), halogenated solvents, and petroleum wastes. Generally, SCE is not used in the treatment of soils contaminated with inorganic compounds, although the processes are used extensively in the nonferrous minerals processing industry to separate and recover metals such as copper, nickel, uranium, and zinc. SCE processes employ several distinct process concepts and a wide range of solvents to separate contaminants from soils, sludges, sediments, and wastewaters. The processes are based on well-established scientific principles (see Section 3.2, below).

The following SCE processes were selected for full evaluation:

- B.E.S.T. Process;
- CF Systems;
- Carver-Greenfield Process;
- Extraksol Process;
- Low Energy Extraction Process (LEEP®);
- NuKEM Development (NKD) Process; and
- Soil Restoration Unit (S.R.U.).

Five of the processes employ solvents at standard pressure and temperature. One process, CF Systems, employs gaseous solvents at or near the critical pressure and temperature and another, B.E.S.T., employs critical solution temperature solvents. See table 3.1 (on page 3.3) for a summary of design information concerning each process.

The following are the emerging process technologies demonstrated at the bench scale that are briefly addressed in Appendix B:

- Chemical Waste Management Adiabatic Process for the Extraction of Sludges (APES);
- Phoenix Miljø Contex Process;
- Henkel Liquid Ion Exchange (LIX) Process;
- Davy R & D Resin-in-Pulp/Carbon-in-Pulp (RIP/CIP) Process; and
- Martin Marietta Soilex Process.

Three additional processes were identified in the literature search for possible inclusion in this monograph. However, due to lack of available process data, lack of response from the process vendor, and/or indication

**Table 3.1**
Design Information

| Process | Solvent(s) | Extractor(s) | Solvent Recovery | Preferred Media |
|---|---|---|---|---|
| B.E.S.T. | Triethylamine | Column or batch | Evaporator and steam stripper | Soils, sludges, and sediments |
| CF Systems | Propane/butane carbon dioxide | Column | Differential pressure | Slurried solids |
| Carver - Greenfield | Food-grade isoparaffinic oil(Isopar-L), Iso - Octanol | Evaporation and solvent extraction stages | Heating, distillation, and steam stripping | High water content wastes |
| Extraksol | Proprietary | Batch | Distillation | Soils, sludges, and sediments< 30% moisture< 40% clay |
| LEEP® | Acetone, others Proprietary | Multistage countercurrent continuous | Liquid - liquid exchange dryer, and distillation | Soils, sludges,and sediments |
| NKD | Proprietary | Mixer/settler or column | Distillation | Soils, sludges, and oily wastewater |
| S.R.U. | Proprietary | Countercurrent continuous | Distillation | Soils, sludges, and sediments< 20% moisture low clay |

that process development efforts had ceased, they were not included. These processes are:

- Environmental Canada Soil Treatment Process;
- Extrapure Process by EM&C; and
- Solvent Extract for Dredged Sludges by SRE, Inc.

## 3.2 *Scientific Basis*

Solvent/chemical extraction effects the preferential separation of one or more constituents from one phase into a second phase. SCE is differentiated form soil washing by the use of concentrated non-aqueous reagents. In theory, SCE can achieve higher removal efficiencies and increased contaminant concentration as compared to soil washing, because both chemical and physical processes are employed.

In a conventional liquid-liquid contacting system, the solution to be treated is called the feed, the material to be extracted is called the solute, and the liquid selected to separate the solute is called the solvent. The solvent-rich, solute-laden product is called the extract, and the residual of the feed stream (from which solute has been removed), the raffinate. The solute concentrations in two contacting liquid phases, corresponding to equal chemical or thermodynamic potentials, define the equilibrium state. The ratio of these concentrations is the "equilibrium distribution coefficient." This is a measure of the best separation or solute removal that can be effected.

In classical chemical engineering terms, SCE is the term applied to the transfer that occurs between two liquid phases. The two liquid phases can be immiscible or partially miscible. Maximum separation of contaminants is effected under the following conditions:

- the substance to be transferred between the source and the solvent phases, the "solute," is much more soluble in the solvent phase;
- the solvent phase is completely immiscible with the feed; and
- the solvent has a substantially different specific gravity from that of the feed.

If solute transfer takes place from a solid substrate into a liquid phase, the action is termed leaching. Common examples of leaching are the recovery of a metal (solute) from metal ore (substrate) by treatment with strong acid and loss of fertilizer from crop land by runoff and percolation of incident rainfall.

Although water is a principal leaching agent or solvent, the separation of organic and inorganic contaminants from soil utilizing nonaqueous liquids is also called leaching. SCE includes the controlled leaching of contaminants from soils, sediments, and solid wastes through use of organic solvents or nonaqueous liquids.

Where liquid-liquid miscibility is very poor (i.e., <1,000 mg/L or 0.1 wt.%) or merely partial, maximum possible transfer from a solid substrate is a function of relative solubilities and the equilibrium distribution coefficient. Where the solute is bound to a solid substrate, solubility of the solute in the solvent is balanced by low-energy sorptive binding, high-energy chemisorption, or incorporation in the solid matrix. The chemical potential of the substrate in the solid phase is a function of substrate-solid interactions: weak van der Waals induced dipole forces versus strong hydrogen, covalent, and electrostatic bonds. The stronger the interactive binding, the poorer the equilibrium distribution coefficient.

A useful concept in explaining liquid-liquid or solid-liquid solvent extraction processes is the "contact stage" or "equilibrium stage." Feed (the liquid or solid substrate containing the contaminant to be removed), and solvent are combined in a mixer or contactor and allowed to approach equilibrium; subsequently, the mixed phases are settled in order to separate the extract and raffinate phases. The combination of mixing and settling constitutes a single contact stage. In turn, stages can be combined in process trains. Partially-purified feed can be repeatedly brought into contact with fresh solvent, reaching new equilibrium states at successively lower solute concentrations. This design is referred to as cross-flow staging. Alternatively, stages that approach equilibrium can be arranged in the counterflow mode described in Section 3.3, below. The final feed-side residue (effluent) stage approaches equilibrium with solute-lean solvent.

The capacity of a solvent to separate substrate x from a weakly or partially soluble liquid or solid, its selectivity, is given by the following equation:

$$Selectivity = \frac{(MassFraction\ \ x\ \ in\ \ E)/(MassFraction\ \ A\ \ in\ \ E)}{(MassFraction\ \ x\ \ in\ \ R)/(MassFraction\ \ A\ \ in\ \ R)}$$

where:

A = Primary feed stream constituent

E = Solvent-rich phase

R = Residual phase (raffinate) at equilibrium

Selectivity must exceed unity; if it is unity, no separation is possible. If A is water, as it is in oily wastewater, secondary sludge, sediment, or wet soil, selectivity may determine whether the extraction technology is applicable.

The vast majority of SCE processes employ solvents at near-ambient pressures and temperatures during the extraction stage(s). Typical solvents include alkanes, alcohols, ketones, or chlorinated solvents, used either singly or in combination.

Solvent extraction can occur under three processing approaches. The most common approach employs two phases in contact at ambient (normal) pressure and temperature, in which solute (contaminant) is exchanged between a solid or liquid substrate and a liquid solvent (at standard pressure and temperature). High pressure and moderately-elevated temperatures can be used to create efficient, dense, solvents or supercritical fluids from substances that are gases at moderate conditions (near-critical fluids). In some instances, temperatures can be increased selectively to enhance solute transfer to a solvent phase (critical solution temperature).

In another approach, near-critical fluid/liquefied gas processes use butane, isobutane, propane, carbon dioxide, or other gases liquefied under pressure at or near ambient temperature during extraction. These processes take advantage of special properties of gases when they are near their critical temperature and pressure (thermodynamic critical point). At this point, the liquid and vapor phases of the solvent, in equilibrium, become identical, forming a single phase. A fluid near its critical point exhibits the viscosity and diffusivity of a gas, while also exhibiting the solvent characteristics of a liquid. Under these conditions, the solvent can very effectively penetrate the solid matrix and mobilize organic contaminants.

Finally, critical solution temperature SCE processes use solvents in which solubility can be varied over the process operating temperature range. These processes use liquid-liquid extraction at two different temperatures. At the lower operating temperatures, the solvents are miscible, while at the upper operating temperatures, the two solvents are completely immiscible. In these processes, solvent recovery often consists of numerous unit operations.

## 3.3 Process Description

Solvent/chemical extraction processes operate in either a batch or continuous mode, and all employ quite similar unit operations. See figure 3.1 for a diagram of a simplified process flow. The following are major unit operations:

- feed preparation;
- extraction;

**Figure 3.1**

Solvent/Chemical Extraction — Simplified Process Flow

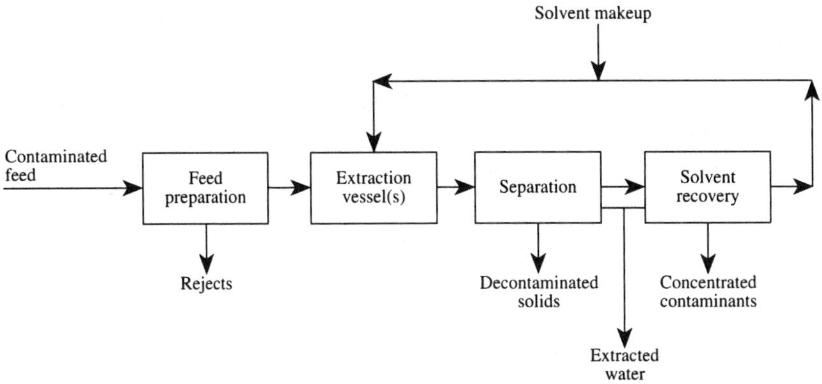


- solids and solvent separation; and
- solvent recovery.

Contaminated soils, sludges, or sediments are excavated and enter the feed preparation system, where they may be screened, crushed, dewatered, and/or slurried, depending on the particular SCE process being employed. Chemical conditioning, such as pH adjustment, may be necessary to assure successful extraction.

The prepared feed is then transferred to the extraction vessel(s) where it is mixed with the extraction solvent(s). Extraction is carried out in either a batch or continuous mode in one or a series of vessels. Selection of the extraction solvent(s), the solvent-to-solids ratio, the extraction contact time, and the number of extraction stages depends upon the specific contaminant and nature of the feed. These parameters are typically determined during treatability studies. Important solvent characteristics include relative solubility of the solute, immiscibility with the feed, specific gravity, toxicity, flammability, physical properties, chemical reactivity, ease of recovery for recycle, and cost.

Feed and solvent streams can enter a continuous contact system in parallel flow or counterflow configurations. In the counterflow arrangement, relatively clean solvent contacts solute-lean raffinate, while feed contacts solute-rich extract. This permits the approach to equilibrium by both end-state pairs. The effluent streams in a parallel-flow configuration can also be caused to approach equilibrium through use of multiple extraction stages. In both cases, the solvent is selected to maximize the solute distribution coefficient.

Following extraction, the decontaminated solids are separated from the contaminant-loaded extraction solvent(s). This may be effected in the extraction vessels or separately through gravity separation, filtration, centrifugation, pressure reduction, or distillation. Residual solids will normally be subjected to multiple washes in order to achieve cleanup goals. The separated solids may retain some solvent, which is removed through distillation, desorption, or an additional extraction step.

The contaminant-laden solvent, along with solvent vapors removed during desorption or raffinate stripping of the decontaminated solids, is transferred to a solvent recovery system. Solvent recovery is effected through distillation, steam stripping, pressure reduction, or phase separation. The recovered solvent is typically recycled back to the beginning of the SCE process, and the concentrated contaminants are removed for further treatment. The extract containing the concentrated contaminants typically requires further treatment before disposal or recovery.

## 3.4 Status of Development

Solvent/chemical extraction systems evaluated here have undergone extensive testing at the bench- and pilot-scale levels, and several commercial-scale systems are in operation. Table 3.2 (on page 3.9), summarizes the status of development of these systems, which is discussed in subsequent sections devoted to each.

## 3.5 Design Data

Available design data vary with process vendors. The most detailed information is proprietary. Five of the systems have been used commercially,

**Table 3.2**
Status of Development

| Process | Quantity for Treatability Study | Pilot Unit Capacity | Commercial Unit Capacity | Comments |
|---|---|---|---|---|
| B.E.S.T. | 20 L (5 gal) | 45 kg/day (100 lb/day) | 90 tonne/day (100 ton/day) | Third generation commercial system designed. |
| CF Systems | 2 to 4 kg (5 to 10 lb) for bench scale. | 45 to 90 kg/day (100 to 200 lb/day) | Sludge/solids: 9 to 900 tonne/day (10 to 1,000 ton/day)<br>Wastewater: 20 to 550 L/min (5 to 150 gal/min) | Variety of commercial scale units available. |
| Carver-Greenfield | 2 to 4 kg (5 to 10 lb) for bench scale. 40 to 400 kg (100 to 1,000 lb) for pilot plant. | 45 kg/hr (100 lb/hr) | Representative range: 4.5 to 45 tonne/day (5 to 50 ton/day) | Stated capacity of 18 installed commercial units (including drying units): 0.3 to 420 dry ton/day. |
| Extraksol | 25 L (6 gal) | 0.9 tonne/hr (1 ton/hr) | 0.9 tonne/hr (1 ton/hr) | A 5.5 to 9 tonne/hr (6 to 10 ton/hr) unit is constructed. |
| LEEP® | 3 to 4 kg (7 to 15 lb) | 90 kg/hr (200 lb/hr) | 9 tonne/hr (10 ton/hr) 7 tonne/hr (7.7 ton/hr (dry)) mobile plant Planned goal 13 tonne/hr (14.3 ton/hr (dry)) stationary plant | Commercial-scale unit designed. |
| NKD Soil Treatment | Bench: 30 grams (1 oz) | None | None | |
| NKD Refinery Waste | | 480 L/day (126 gal/day) | None | |
| S.R.U. | 6 to 100 kg (15 to 250 lb) | | 1.8 tonne/hr (2 ton/hr) | Mobile unit constructed. |

and two (CF Systems and Carver-Greenfield) have undergone US EPA SITE Program evaluations. For five processes, US EPA technology profiles have been completed, and US EPA technology evaluation reports are available for three (US EPA 1992g; 1989; 1988; 1992a). Sizing and scale-up data are available for all of the processes. Specific design data for each process are presented in the section devoted to that process.

## 3.6 Pre- and Posttreatment

The seven SCE processes addressed can treat a wide variety of materials; but all require some feed pretreatment. Pretreatment generally consists of physical sizing and, in some cases, chemical conditioning of materials before they are fed to the extraction vessel. See table 3.3.

Soils must be screened after excavation to remove oversized material, which may then be rejected or crushed and treated, depending on the feed material and the type of contamination. Some processes may require the addition of water to a minimum level, whereas others may require solids and liquids separation.

The products of the SCE process include decontaminated solids, a concentrated contaminants fraction, wastewater, and contaminated solvents. Each of these streams may require additional treatment depending on required cleanup levels, the nature of the contaminants, and the final disposition of the stream. The decontaminated solids may contain trace amounts of solvent, as well as the original contaminants. Under some process conditions, naturally-occurring organic substances will be extracted and bacteria

**Table 3.3**
Summary of Pretreatment Requirements

| Process | Pretreatment Requirements |
|---|---|
| B.E.S.T. | Screen to 6 mm (0.25 in.). pH adjustment to 10.5 to 11. |
| CF Systems | Screen to 9 mm (0.19 in.). pH adjustment to between 6 and 10.<br>Viscosity adjustment to <5,000 cP. Temperature to 15°- 50°C (60°-120°F) range. |
| Carver-Greenfield | Treats wastes up to 99% water, but staged process may require evaporative stages prior to extraction stages for high water content wastes.<br>Screen particles to 6 mm (0.25 in.). |
| Extraksol | Screen porous material to 51 mm (2 in.).<br>Dewatering to <30% moisture content. |
| LEEP® | Removal of debris (>203 mm (>8 in.)).<br>Crush and screen to 13 to 25 mm (0.5 to 1 in.).<br>Free water removed by filtration or centrifugation. |
| NKD Soil Treatment | Screen to remove debris.<br>Reduce size and/or screen to <50 mm (2 in.) particle size. |
| NKD Refinery Waste | Feed cannot contain >60 mesh particles. |
| S.R.U. | Screen porous material to 7.6 cm (3 in.).<br>Dewater to <20% moisture content.<br>Solvent premix for high concentration wastes. |

will be killed. This will render the soils sterile, possibly requiring additional treatment to reestablish beneficial bacteria.

The concentrated contaminant stream will contain the organic contaminants, oil and grease, some naturally-occurring organic substances, and small amounts of the extraction solvent(s). The concentration of organic contaminants may be on the order of several hundred to ten thousand times that in the contaminated feed. This low-volume, highly-concentrated stream will most likely have to be destroyed by incineration, wet oxidation, or other destructive process. In some cases, depending on the contaminants present, it may be possible to recover this stream for recycling/reuse.

When moisture-containing feed materials are processed, a solvent and water mixture is often generated. This mixture is normally separated through distillation, which will produce a water stream containing low-volatility, water-soluble contaminants, as well as small amounts of residual solvent. The quantity and quality of water produced will vary from application to application. Additional treatment will depend on the ultimate use of the water. In some cases, it may be possible to discharge it directly to a publicly owned treatment works (POTW).

Pre- and posttreatment requirements specific to each process are discussed in the section devoted to that process.

## 3.7 Environmental Impact

Each of the processes addressed can potentially impact air, water, and land. Air emissions can result from process leaks and/or from purposeful venting of gases. Vented air can be treated by carbon adsorption or other systems. The majority of the chemical solvent extraction systems reviewed use nitrogen blankets to minimize emissions and minimize potential explosive hazards. Because some of the solvents used are proprietary, comprehensive information on potential air emissions problems was not available. It is known, however, that odor problems have arisen in some tests. The B.E.S.T. system, for example, uses triethylamine (TEA), which has a strong fishy/ammonia odor. During full-scale operation, leaking seals caused excessive emissions, which resulted in Level B protection (OSHA) being specified for operating personnel. This level of protection is not expected for a commercial operation.

Residues from each process might impact both water and land. Often, coarse fractions cannot be treated because particles are oversize. These fractions may be crushed for either processing in solvent extraction systems or they may be treated by other methods. Increased handling of contaminated materials, however, especially materials contaminated with volatile organic compounds, may result in excessive fugitive emissions.

Solvent extraction produces water and solid fractions that may adversely affect water and land. Water is separated from the treatment solvent in each process. The residual wastewater will generally contain residual organic contaminants and extraction fluid, as well as soluble heavy metals present in the feed material. Treatment will be required before discharge to a water course and may be required before discharge to a POTW.

Residual solids are separated from the treatment solvent once extraction is complete. If the feed was a hazardous waste, the treated soil may require disposal as a hazardous waste. Alternatively, after being treated, the soil may be classified as non-hazardous.

In some instances, treated soil may exhibit higher concentrations of metals than the feedstock because of the effects of volume reduction. The actual leachability of the metals, however, may not increase, and therefore, the solids may be considered clean. Even if elevated levels of inorganics remain, the removal of organic contaminants may facilitate additional treatment, such as stabilization.

If the solids are clean, they may be backfilled on site or used for other purposes. During treatment, solids are dried. Several systems add steam or water to the treated solids to reduce dust. Other systems discharge dry, fine, dusty materials, which may be treated with additives to control dust. In some instances, however, the cleaned solids fraction may contain residues of the original organic contaminants and the extraction fluid and may require further treatment. If heavy metals are present in the raw feed material, they may become concentrated in the cleaned solids. This residual may be classified as a Resource Conservation and Recovery Act (RCRA) hazardous waste, subject to land disposal restrictions and, therefore, require stabilization or other treatment.

The final residues of concern are the solvent and the extracted organic contaminants. In most systems, the solvent is recovered and recycled. The concentrated contaminant fraction may be classified as a hazardous waste and, therefore, be subject to regulation under RCRA. Such a waste stream must be destroyed through incineration or other appropriate means. Once treatment is completed, the solvent will generally require disposal as a hazardous waste. Again, because some of the solvents are proprietary, comprehensive disposal issues for the solvents cannot be examined here.

Specific environmental impacts of each process are addressed in the section devoted to that process.

# 3.8 *Basic Extractive Sludge Treatment Process*

## 3.8.1 Description

Resources Conservation Company (RCC) of Bellevue, Washington, developed the patented B.E.S.T. Process, in which the unique miscibility properties of certain amine solvents are applied to separate sludges or oil-contaminated solids into their oil, water, and solids fractions (Robbins 1990; Tose 1987; Weimer 1989). Triethylamine (TEA) is the organic solvent used in this process. At temperatures below 18°C (64°F), TEA is fully miscible with water, and above this temperature, only slightly miscible. Therefore, cold TEA can be used to dewater solids, and, at the same time, organic contaminants are partially removed. The remaining organic contaminants are removed by TEA heated to temperatures above 55°C (130°F) and below 80°C (176°F). Within this temperature range, TEA is liquid and is not miscible with water (Weimer 1989).

In the B.E.S.T. Process (figure 3.2 on page 3.14), contaminated solids are loaded into a washer/dryer unit, a steam-jacketed vessel with a horizontal mixer shaft, to 25% of its capacity. A predetermined amount of caustic is added to increase the final mixture pH to between 10.5 and 11. The vessel is sealed and purged with nitrogen gas to remove residual oxygen and to provide a nonexplosive environment for the washing and drying steps.

**Figure 3.2**
B.E.S.T. Process Schematic

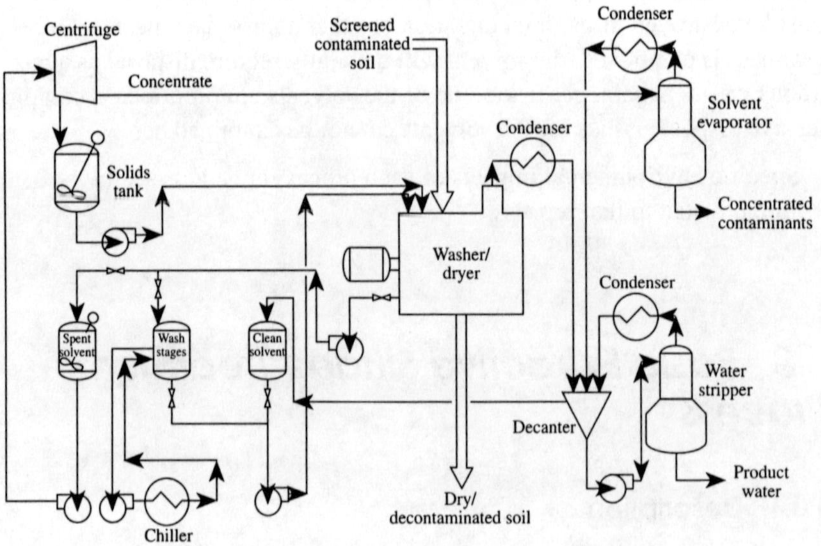


Chilled solvent (<6°C (<43°F)) is added to fill the vessel, resulting in a solvent-to-soil ratio of 3:1. The extraction phase begins when the mixer shaft motor is activated. Paddles, affixed to the shaft, sweep the inside diameter of the vessel. This agitation mixes the solvent and solids. Mixing time varies from 5 to 15 minutes per extraction.

At the end of each extraction step, the mixer is stopped and the solids are allowed to settle. Since TEA has a low specific gravity (0.73 @ 20°C (70°F)), solids, which would normally stay in suspension in water, will settle out. The TEA, water, and oil mixture is decanted and pumped to a centrifuge. The centrifuge separates any residual solids from the mixture and returns the solids to the washer/dryer for additional extractions. The concentrate is heated in an evaporator to about 77°C (170°F). At this temperature, the TEA/water azeotrope will evaporate, leaving the oil phase, which is removed from the bottom of the evaporator. The TEA/water vapors are condensed and flow into a decanter. Since TEA and water are immiscible at temperatures above 55°C (130°F), the condensed TEA and water separate into two distinct phases in the decanter. Water is pumped from the decanter to a steam stripper. The steam stripper volatilizes residual TEA and discharges an aqueous stream. Gaseous TEA, discharged from the top of the steam stripper, is condensed and collected in the decanter, and the condensate serves as the feed to the steam stripper. Decanted TEA is recycled for additional extractions.

Subsequent extractions are carried out at temperatures above 55°C (130°F). At this temperature, the solubility of organic contaminants in TEA increases, thus increasing removal efficiencies. At the completion of the extractions, steam is injected into the washer/dryer's jacket, and the mixing shaft is rotated. As the temperature of the washer/dryer increases, TEA volatilizes, thereby producing a dry solids product. Live steam may be injected directly into the washer/dryer near the end of the drying cycle to return moisture to the solids. This measure helps control fugitive dust when the solids are discharged.

### 3.8.2 Status of Development

RCC has developed a B.E.S.T. Process treatability test, performed in the company's laboratory. The test simulates the process steps employed by the full-scale and pilot-plant units and is used to determine the amount of caustic required for pH adjustment. It is also used to determine the total number of extraction stages required to achieve cleanup goals.

For this test, chilled TEA is mixed with a measured quantity of solids. The solids are permitted to settle, the TEA is decanted, and the solids are centrifuged. The solids may be sampled at the completion of each extraction for analysis. Alternatively, if high removal efficiencies (>95%) must be achieved to demonstrate effectiveness, solids sampling may be initiated after the completion of several extractions (Robbins 1990).

RCC has constructed a 44-kg/day (100 lb/day) B.E.S.T. pilot plant. The skid-mounted module contains all the unit processes and is easily transported on highways. The pilot plant was used for treatability testing on PCB-contaminated soils at spill sites in Ohio and New York (Robbins 1990). Additional SITE demonstration testing of this system has been completed (Meckes et al. 1992; US EPA 1993).

### 3.8.3 Design Data

As explained in Subsection 3.8.1, the B.E.S.T. Process is designed to capitalize upon the unique properties of the organic solvent TEA. When contaminated solids are mixed with chilled TEA, the water and organic contaminants will be extracted. Once dewatering of the solids has been completed, additional extractions are conducted at temperatures above 55°C (130°F). At this temperature, the solubility of organic contaminants in TEA increases, thereby enhancing the capacity for removal of organic wastes.

According to Tose (1987), design throughput for the full-scale B.E.S.T. transportable system is 91 tonne/day (100 ton/day). Maximum throughput during cleanup activities at the General Refining Site, however, was shown to be 64 tonne/day (72 ton/day). Nominal composition of the sludge treated at this site was 70% water, 20% solids, and 10% oil. During operation, the composition of the sludge ranged from 60% to 100% water, 0% to 40% oil, and 2% to 30% solids (US EPA 1988). Throughput was found to vary with the feed material's composition. Furthermore, the need for high contaminant removal efficiencies associated with contaminant levels may have necessitated multiple extraction stages. This would also reduce rated throughput.

The system used at the General Refining Site is designed to accept pumpable fluids. Waste streams that are not pumpable must be modified to accommodate the pumps feeding the extraction column.

The B.E.S.T. pilot plant uses a washer/dryer system (Robbins 1990; Weimer 1989). This system represents RCC's current approach to the B.E.S.T. Process. Both pumpable or nonpumpable solids may be loaded into the washer/dryer, where all operations involving the solids take place. This practice minimizes solids handling requirements and results in a batch-feed operation. Since dewatering of solids can be effected by use of chilled TEA and separation of TEA/water mixtures can be induced by increasing the mixture temperature above 55°C (130°F), high moisture content solids, such as sludges and sediments, may be dewatered with a low expenditure of energy.

### 3.8.4 Pre- and Posttreatment

The full-scale B.E.S.T. systems can treat pumpable waste flows with a particle size < 2.5 cm (1 in.) in diameter. The B.E.S.T. pilot plant can treat pumpable and non pumpable solids with a particle size < 6 mm (0.25 in.). Material classifiers, such as shredders and screens, may be required to reduce the waste to these particle sizes. According to RCC, future production models will accommodate solids with a diameter ≤ 2.5 cm (1 in.). In order to maintain TEA in a nonionized state, the pH of the final extraction mixture must be between 10.5 and 11. The amount of caustic to be added to the extractor is determined by treatability tests (Robbins 1990; Tose 1987; Weimer 1989).

Solids discharged from the B.E.S.T. Process are dry and oil free with a pH in the range of the final extraction mixture. Some residual solvent and/or organic contamination may remain. If the original waste stream contained inorganic contaminants, such as metals, they will remain with the solid fraction. Solutions containing concentrations up to 200 mg/kg TEA in water are known to biodegrade. No available data, however, document degradation of TEA in soils (Meckes et al. 1992). When offered for disposal, concentrated TEA is considered to be a hazardous waste since it meets the definition for ignitability in 40CFR261.21. However, it is unlikely that residual TEA concentrations of <200 mg/kg in soils will meet the regulatory definition of "ignitability".

Solids meeting cleanup criteria may be returned to the land for reuse. Since beneficial organic matter may also be removed by the extraction process, augmentation of the treated solids with organic substrate may be necessary to promote degradation of residual solvent or support vegetative growth. Depending on the extent of metal or other inorganic contamination, the solvent-extracted solids may need to be treated by some other technique, such as stabilization/solidification or soil washing (Meckes et al. 1992; US EPA 1990c).

The physical properties of the treated solids are similar to the feed material. In many cases, they can be compacted and backfilled (after the addition of water) as is. While in other instances, they may require the addition of stabilizing materials to achieve acceptable compaction properties.

Water produced through this process should be analyzed for the contaminant(s) of concern and residual solvent. Low concentrations of organic contaminants may remain with the water fraction, although, the water may be acceptable for discharge to a POTW. Liquid phase carbon adsorption may be used to produce an aqueous discharge free of organic contaminants.

Organic contaminants are concentrated in the oil fraction. If chlorinated hydrocarbons, such as PCBs and many pesticides, are among the contaminants of concern, the treatment options are incineration or chemical dehalogenation. If chlorinated hydrocarbons are not among the contaminants of concern, the concentrated waste stream may be recycled.

### 3.8.5 Operational Considerations

The 91 tonne/day-transportable unit, excluding support equipment, occupies an area of approximately 335 $m^2$ (3,600 $ft^2$) (US EPA 1988). Support equipment consists of material classifiers, makeup solvent tanks, caustic tanks, and water treatment equipment. This unit is designed to treat pumpable solids only. Therefore, some modification of the waste stream may be required. RCC's B.E.S.T. pilot plant uses a washer/dryer system that will treat pumpable or nonpumpable solids.

Triethylamine, having an open cup flash point of -4°C (25°F), is a highly flammable solvent. Transfer of solvent from storage to process tanks must be effected with nonsparking equipment. Chapter 5 of the National Fire Protection Association (NFPA) standard states that a restricted access zone must extend to a 15-m (50 ft) radius around the extraction plant and a control zone must extend from the 15-m (50 ft) line to a radius of 30 m (100 ft) from the extraction plant (NFPA 1990). Therefore, siting the system at some locations may be difficult.

### 3.8.6 Environmental Impacts

Triethylamine has a strong ammonia-like odor. During full-scale operation, seals on a centrifuge developed leaks, released low concentrations of TEA, and caused an offensive odor (US EPA 1988). Newer washer/dryer designs do not use the same type of system for product dewatering; vapors discharge through activated carbon filters, which mitigate the odor. Wastewater generated via this process may contain low concentrations of organic and inorganic contaminants and require treatment before discharge.

# 3.9 CF Systems

## 3.9.1 Description

The CF Systems Corporation (CFS) of Woburn, Massachusetts, and CF Technologies Inc. (CFT) of Hyde Park, Massachusetts, specialize in the development and application of supercritical fluid and liquefied gas extraction processes for chemical production and hazardous waste treatment. Liquefied gases are used as extracting solvents to separate organic solutes and sorbates from wastewater, sludge, and contaminated soil. Target contaminants include hydrocarbons (benzene, toluene, xylene (BTX) and constituents of gasoline), oil and grease, partially-oxidized hydrocarbons (phenol, alcohols, fatty acids, acetone, etc.), and chlorinated species (PCBs and dichloroethane). Carbon dioxide ($CO_2$) is generally used for aqueous solutions; propane is often selected for sediments, sludges, and soils. In selecting the solvent, the solubility of $CO_2$ in water and the effects on pH and soluble inorganic salt content must be considered. Propane is a volatile, flammable, hydrocarbon that can constitute a fire and explosion hazard in the event of system malfunction.

Figure 3.3 (on page 3.20) is a simplified diagram of a one-stage solvent extraction process employing liquefied propane. Contaminated sediments are fed top down into a high-pressure contactor. Compressed liquefied propane at 20°C (70°F) passes upward, counter to the solids, and dissolves organic matter. Clean sediment (raffinate) is removed from the contactor. A solution of organic contaminants in propane is passed to a separator via a pressure-reducing valve. Propane is vaporized, recompressed, and recycled to the contactor as fresh solvent. Contaminants and natural organic matter are removed from the separation vessel and recovered for disposal or reuse.

The process has seven basic operating steps. Initially, slurried sludge is fed to a stirred-tank extractor (raw sludge may require pretreatment to eliminate oversized material or to modify chemical characteristics, such as pH). Propane is compressed to operating pressure, condensed, and fed to the extraction vessel to dissolve oil in the sludge feed. A mixed stream is taken from the contactor to a decanter, in which gravity separation of the heavier water and solids fraction and the lighter propane and oil fraction occurs. Water and treated solids are removed from the decanter; the solids are dewatered and the final filter cake is removed to landfills.

Propane and oil pass to a solvent-recovery still. The distillation tower operates at a reduced pressure. The reboiler is heated with recompressed propane vapor. Oil, collected as still bottoms, is recycled to the refinery, and propane is recycled as fresh solvent.

This one-step mixer/settler system is actually operated as a multiple-stage process. The number of stages must be suitable to achieve best demonstrated available technology (BDAT) standards for refinery hazardous wastes (K048-K052) prescribed by US EPA. The number of stages required (typically 2 to 5) is dependent on the feed matrix and type and level of contaminants present. The treated oil and solids raffinate stream from the commercial unit is claimed to conform to BDAT standards for sixteen specific volatile and semivolatile organic compounds.

The same concept can be applied with supercritical carbon dioxide or liquefied light hydrocarbon gas mixtures as the solvent. These modified processes have been evaluated at bench and pilot scales.

**Figure 3.3**

CF Systems Solvent Extraction Unit Process Diagram

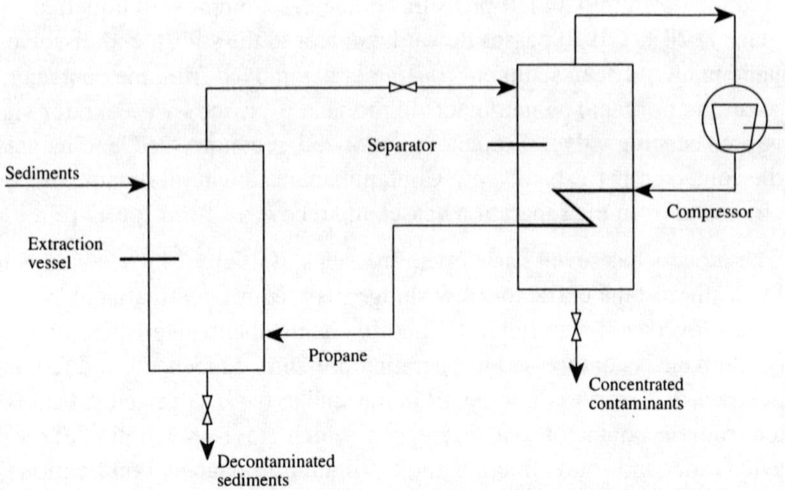


US EPA 1990a

### 3.9.2 Status of Development

Supercritical $CO_2$ has been used by CFS in treating contaminated groundwater and soils, wastewaters, and sludges. Removal of volatile and semivolatile organics has been demonstrated at bench scale. Liquefied propane has been used to extract organic contaminants from refinery sludges, separator sludges, and soils. Bench operations employ a single-contact stage that can be used repetitively to aid in estimating the number of pilot- and full-scale contact stages necessary to attain the desired extraction efficiency. Results of bench-scale studies approach classic equilibrium thermodynamic measurements. In general, small-scale studies have suggested that the use of liquefied gas in extracting organic contaminants from semisolids is somewhat less efficient than in extracting them from aqueous wastes.

At the pilot scale, CFS has demonstrated a trailer-mounted unit, the Pit Cleanup Unit (PCU), in treating soils, sludges, and semisolids with liquefied propane and a propane/butane mixture. This light hydrocarbon supercritical extraction unit uses two stages of countercurrent extraction with solid-liquid separation between extractors. Feed slurry is passed through a basket filter to remove particles larger than 3 mm (0.12 in.) in diameter. An agitator provides mixing action between contactor and decanter pairs. Two immiscible phases develop in Decanter 1; solids and water settled in this unit are pumped to Extractor 2. Decanter overflow, containing extracted organic substances, propane, butane, and fine solids, is filtered and sent to a solvent recovery still.

In the primary demonstration of the PCU, at the New Bedford, Massachusetts, Superfund Site, the Corps of Engineers dredged and drummed sediments contaminated with concentrations of PCBs as great as 2,600 ppm (US EPA 1990a). In addition, CFS has used the PCU to treat soils and sludges at nine refineries, petrochemical plants, and treatment, storage, and disposal (TSD) facilities. The PCU has also been used to treat a 1:1 mixture of American Petroleum Institute (API) separator sludge and dissolved air flotation (DAF) sludges, material from a clay pit, ditch skimmer sludge, and tank bottoms.

The propane extraction process has been deployed to treat sludge at the Texaco Star Enterprise petroleum refinery in Port Arthur, Texas. Treatment plant capacity is approximately 23 tonne/day (25 ton/day). The solvent is liquefied propane at pressures of about 2.1 MPa (300 psig) and temperatures of 32° to 49°C (90° to 120°F).

### 3.9.3 Design Data

Analysis of commercial liquefied propane extraction of refinery sludge indicates that raffinate conforms to BDAT standards for volatile and semivolatile organic compounds. See table 3.4 for a summary of results of testing performed on a commercial system. Extraction efficiencies for the class of contaminants under consideration varied from 80% to approximately 100% for liquefied propane extraction using typically 1 to 5 extraction stages.

**Table 3.4**
CF Systems - Commercial System

| Compound | Concentration (ppm) | |
|---|---|---|
| | Typical Treated Solids | BDAT Standard |
| Benzene | < Detection Limit | 14 |
| Toluene | < Detection Limit | 14 |
| Total Xylene | 3.8 | 22 |
| Naphthalene | 10.7 | 42 |
| Benzo(a)pyrene | < Detection Limit | 12 |

Pilot-scale studies of liquefied light hydrocarbon gases have been performed using the PCU on a number of soils and wastes, including: (1) a contaminated soil from the United Creosote Superfund Site, Conroe, Texas, (2) a 1:1 mixture of API separator sludge and DAF sludge, (3) clay-pit residues (Port Arthur, Texas), and (4) ditch-skimmer sludge. Selected data from three of the studies presented in table 3.5 (on page 3.23) reflect the attainment of BDAT standards and concentrations below detection limits after treatment. In pilot-scale studies of supercritical extraction of sediments contaminated with PCBs performed as part of the SITE Program at New Bedford, Massachusetts (see Subsection 3.9.2) the solvent was a 70:30 mix of propane/butane. Three series of evaluations were conducted with the two-stage operating systems. A series consisting of a number of sequential passes of sediments contaminated with PCBs in concentrations of 288 to 2,575 ppm demonstrated great variations in distribution coefficients

**Table 3.5**
CF Systems - Pilot Plant Data

| Compound | Concentration (mg/kg) | | | | | |
|---|---|---|---|---|---|---|
| | United Creosote Soil | | Mixed Sludge | | Clay Pit Residue | |
| | Feed | Treated Soil | Feed | Treated Soil | Feed | Treated Soil |
| Water (wt %) | | | 53.3 | 34.1 | 60.5 | |
| Solids (wt %) | | | 38.0 | 65.9 | 22.3 | |
| Oil and grease (wt %) | | | 8.7 | 0.012 | 17.2 | 1.9 |
| Benzene (ppm) | | | 11.9 | 0.13 | 9.6 | <0.1 |
| Toluene (ppm) | | | 88.5 | 0.5 | 16 | <0.1 |
| Naphthalene (ppm) | 140 | 1.5 | 47.8 | | 210 | <5.3 |
| Pyrene (ppm) | 360 | 11 | 12.5 | 0.3 | | |

and solids retention. Removal efficiencies were as high as 90%; extraction efficiencies of 60% were generally attained in the first pass of each test series. A mass balance for PCBs was not possible. A greater mass of contaminant was reported for process effluents and for decontamination washes than had apparently been fed to the system. The cause of these anomalies is not known.

In three PCB-contaminated sediment tests, feed slurry flow varied between 2.5 and 6.5 L/min (0.6 and 1.4 gal/min). Supercritical solvent flow varied widely outside the desired range of 3.5 to 7 kg/min (8 to 15 lb/min); thus, the solvent/feed ratio also fluctuated widely.

Bench-scale studies are best conducted in a single-stage, near-equilibrium context. With supercritical $CO_2$, CFS reports very high thermodynamic efficiencies; at concentrations of 0.4 to 520 ppm, extraction efficiencies are reported to vary from 95 to 99.99%. These measurements are equivalent to distribution coefficients as large as $10^5$ for chloroform, dichloroethane, and trichloroethane. The distribution coefficients for Aroclor 1242 and phenol are approximately 20.

## 3.9.4 Pre- and Posttreatment

Pretreatment of wastes in supercritical extraction will be specific to the site and materials to be handled. Experience with sediments from New

Bedford, Massachusetts, indicated a need to remove particles greater than 19 mm (0.75 in.) in order to decrease viscosity and establish a suitable feed temperature. Feed consistency must be relatively homogeneous to provide adequate control of flow rates. Solids removal, water addition, mixing, and storage capacity are important pretreatment considerations. In the New Bedford Harbor SITE demonstration test, using supercritical propane, sieving and screening were important (US EPA 1990a).

Vibrating screens are widely used because of their large capacity and the resulting high efficiency; however, wet or sticky materials tend to blind the screen. Wet screening with sprays is probably appropriate to inhibit blinding. Manual or automated high-pressure water sprays are assumed to be adequate to treat oversized solids. Coarse solids would be disposed of with fine-grained materials treated by CF Systems' supercritical extraction technology. Spray water should be collected and reused.

Posttreatment must be considered for two product streams generated by the supercritical extraction process. The extract contains concentrated organic matter and treated sediments contain water and solids. Extracted contaminants must be contained, handled, stored, and transported off site in a manner conforming with RCRA requirements. The volume of raw sediment would be less than the volume of treated effluent. The wet product must be dewatered, and the water must be returned for feed stream preparation to minimize wastewater treatment requirements.

Pretreatment for the CF Systems SITE demonstration at the New Bedford site consisted of dredging, sediment storage and handling, coarse solids separation, water addition, and temperature adjustment. Posttreatment consisted of dewatering treated sediment, disposal of treated sediments, and off-site disposal of extracted organic matter.

### 3.9.5 Operational Considerations

Foaming in treated sediment and extracted contaminant tanks was evident throughout SITE project experiments (US EPA 1990a). The suspected cause was propane entrainment. Foaming has two adverse effects: (1) extracted contaminant (PCBs) persists in the foam and (2) increased volume of the product stream leads to higher costs and likely contaminant migration. Commercial process designs will probably utilize multiple pressure relief steps to decrease pressure gradually and reduce the potential for propane entrainment.

Retention of solids and oily contaminants in the New Bedford SITE system (see Subsection 3.9.2) influenced the interpretation of test data. The CF Systems pilot unit was operated in a recycle mode to simulate multiple stages. This led to some cross-contamination of recycled, treated sediments. To prevent cross-contamination in operations that involve substantial feed stream changes, oil and contaminant removal with an organic solvent is recommended.

During the SITE tests, there was no acute threat to operator health and safety. Combustible gas meters, used to monitor the test unit, did not indicate any significant leakage of propane, which is highly combustible and poses an explosion threat. High-pressure operations create a greater need for leak testing, monitoring, and pressure vessel design detail than do conventional, domestic propane applications. Background air sampling and personnel monitoring indicated that organic vapor and PCB concentration, if present, were at levels below detection limits. Gases vented from the unit were passed through a carbon canister. They contained minor concentrations of PCBs. The greatest threat to SITE project workers was deemed to be dermal exposure. For personnel handling treated and untreated wastes, Occupational Safety and Health Act (OSHA) Level B personnel protection equipment (PPE) is recommended. Level C PPE is recommended for extraction process operators. The PPE clothing should be designed for the materials handled and the environment of the process.

All electrical equipment must be explosion proof (US EPA 1990a). Potential sources of ignition must be eliminated in the process plant area when propane, butane, or propane and butane mixtures serve as the supercritical extractant. Spark-proof tools must be used in all instances. Solvent recovery is similar to commercial refinery depropanizer operation, which has a long history of safe, efficient use in the international petrochemical industry.

If supercritical $CO_2$ is selected as the solvent, fire and explosion hazards are mitigated. However, $CO_2$ may still present a health hazard in confined areas. Leakage still represents a pathway for losses of extracted contaminant. Total pressure is higher than in the case of hydrocarbons, and this requires careful consideration of high pressure design criteria. System failure at elevated pressure constitutes a source of physical hazard to plant operators. In addition, unexpected pressure release could result in discharge of extracted contaminants before controlled separation and disposal. Again, OSHA Level B and Level C PPE are appropriate because of the hazardous nature of the contaminants present in feed sediment, sludge, or soil.

### 3.9.6 Environmental Impacts

Process systems must be monitored continuously for pressure losses and emissions to the atmosphere. These emissions constitute a fire and/or explosion hazard where light hydrocarbons are used and a source of process instability where supercritical carbon dioxide is used. In both cases, organic contaminants are concentrated in the solvent, and leakage or spillage may release a hazardous vapor or aerosol to the atmosphere. The overall process provides means for contaminant separation, solvent recovery, and solvent repressurization for reuse in waste contacting units. Since the goal is to maximize contaminant concentration in the solvent stream, solvent management is particularly critical. Units in the recycle train must be monitored closely to detect losses of contaminant to the environment.

A hazardous waste feed may lead to a hazardous product. The product may be wastewater requiring further treatment, relatively dry soil, or sediment. After treatment, the product must be evaluated to establish its hazardous characteristics. Treatment after supercritical fluid extraction must be designed to eliminate hazardous characteristics before a product is discharged to the environment. The enriched contaminant stream must be completely destroyed, substantially reduced in toxicity, or stabilized for safe, long-term storage.

## 3.10 Carver-Greenfield Process

### 3.10.1 Description

The Carver-Greenfield Process is a patented drying and solvent extraction treatment process, which is commercially available from the Dehydro-Tech Corporation (DTC). The process separates mixtures into solids, oil, and water, while extracting organics using a carrier oil or solvent (US EPA 1992a, A.1). In instances where heavy metals are complexed by hydrocarbons, some metals may also be removed from the solids (US EPA 1990b). Treatment effectiveness can be increased by adding evaporation and extraction stages.

The process has been variously described as extraction (Trowbridge, Holcombe, and Kollitides 1991), drying (Lau 1991), steam stripping (Hazardous Waste Consultant 1991), and evaporative. It has been characterized by the US EPA both as a "solvent extraction process" (US EPA 1991a) and as "other physical treatment" (US EPA 1991c). Because the main treatment steps involve solvent extraction and water evaporation stages, the process is addressed in this monograph. It should be noted that the carrier solvent may be used in the very first stage or it may be mixed with the waste in later evaporation and extraction stages after some evaporation has already occurred (US EPA 1992a). The solvent aids in maintaining the waste in a slurry state as water is evaporated, as well as serving as a medium for the extraction of organic contaminants. The process consists of the following steps (see figure 3.4 on page 3.28):

1. Pretreatment. Debris is separated from the feed, and if necessary, the feed particles are ground to sizes less than 6 mm (0.25 in.);
2. Feedstock Slurrying (Fluidizing). The feed material is slurried in a fluidizing tank with a carrier oil or solvent to extract indigenous oils and soluble organics. In general, the solvent-to-feedstock waste solids ratio varies from 5:1 to 10:1 by weight. The exact solvent to be used depends on the site, but a hydrocarbon-based solvent with a boiling point around 150°C (300°F) is used for hydrocarbon or organically-contaminated solids (US EPA 1992d). According to DTC, alcohols or food-grade mineral oils are typically used (US EPA 1992d). The product of this stage is a slurry mixture;
3. Evaporation and Solvent Extraction Stages. The water in the slurry is evaporated. In general, two to four multi-effect evaporators are used in commercial systems to evaporate the water (US EPA 1992d). Alternatively, mechanical vapor recompression may be used (Holcombe and Kollitides 1991). According to DTC, lower temperatures can be used, if necessary, by operating under a vacuum (Trowbridge, Holcombe, and Kollitides 1991). For example, the evaporative stages can employ succes-

**Figure 3.4**
Carver-Greenfield Process Schematic

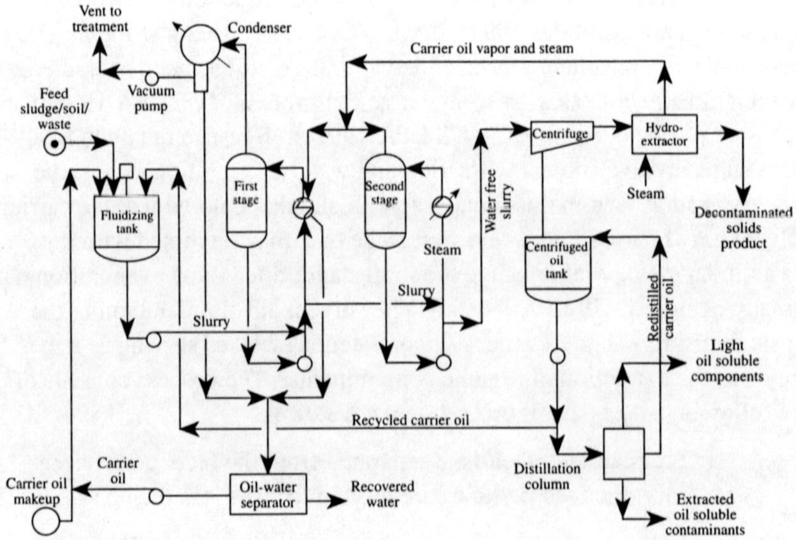


sive boiling chambers, each operating at progressively lower pressures (Environment Today 1991). This allows succeeding chambers to use less energy vaporizing the water (Environment Today 1991). Removal of the water aids in breaking up emulsions, thereby increasing organic extraction. At the same time, steam generated in the evaporation system removes water and volatile compounds from the waste-solvent slurry (NETAC 1991; Hazardous Waste Consultant 1991). The heat also destroys microorganisms. The products of these stages consist of vapors and a water-free slurry of solids in carrier solvent;

4. Condensation and Oil and Water Separation (Vapor Treatment). The vapors from the evaporation step are condensed. The water, carrier oil, and solvent condensate are then sent to an oil and water separator (decanter). The decanting separates any carrier oil and solvent and water-immiscible solvents from the water.

The recovered water contains some residual solvent and low-boiling point water-soluble compounds. The water is, however, generally relatively clean and virtually free of solids and can usually be treated with standard wastewater treatment technologies. Any recovered carrier oil can be recycled to the fluidizing tank. The vent gases can be treated for residual organics by granular activated carbon (US EPA 1992a);

5. Centrifuging (Water-Free Slurry Treatment). The majority of the carrier oil and solvent is separated from the feed solids by centrifuging. The solids may then be reslurried with clean (recirculated) solvent for additional extractions or be directed to desolventization. The concentrate (from each extraction) generally consists of the carrier solvent (with extracted indigenous oil and organics) and approximately 1% fine solids. The centrifuge cake generally consists of 50% solids and 50% solvent with extracted organics;

6. Desolventization of Solids (Treatment of Centrifuge Cake). The solvent is removed from the solids by heating (evaporation) and stripping by countercurrent contacting of the solids with gas (US EPA 1992a). Earlier descriptions referred to this as a "hydroextraction" or "vacuum hydroextraction" step (Holcombe and Kollitides 1991) that heated the centrifuge cake under vacuum (Bress, Greenfield, and Haug) and utilized steam to contact the solids (US EPA 1991a). More recent studies have used nitrogen gas to strip the solids (US EPA 1992g). The resulting offgas is then scrubbed to remove carrier oil/solvent and recirculated. The vent gases can be treated for residual organics by granular activated carbon (US EPA 1992a). Most of the heavy indigenous oils in the centrifuge cake will remain with the solids in the centrifuge cake, rather than evaporate (US EPA 1992d); and

7. Distillation of Carrier Oil/Solvent (Treatment of Concentrate). The used carrier solvent is distilled to recover the carrier oil and solvent and separate the indigenous oils and organics. Products of this step consist of a recovered solvent (substantially free of contaminants), which may be reused, and concentrated streams of light and heavy organics, which may be incinerated or reclaimed.

It should be noted that the above description pertains to the complete Carver-Greenfield Process. Some designs incorporate only some of these stages.

The pilot-scale plant that was used in some recent testing differs from the commercial-scale systems that have been installed to treat other types of wastes. For example, the commercial-scale systems generally operate continuously, while the pilot-plant extraction was conducted on a batch basis. In addition, solvent distillation was not performed during pilot-scale tests (because of the lack of pilot-scale distillation equipment). Therefore, new solvent was used for each extraction. Finally, water evaporation was conducted in a single-effect step.

### 3.10.2 Status of Development

The Carver-Greenfield Process was originally developed by Charles Greenfield in the 1950s at the Carver Press Company laboratory (NETAC 1991). The first commercial plant was installed in 1961 to treat meat rendering wastes (NETAC 1991), and the process has since been operated commercially in the treatment of nonhazardous municipal and industrial wastes (US EPA 1992a). Commercial-scale systems generally treat other types of waste, and consequently, the capacity of the process for treating soils and petroleum sludges has not been proved.

In recent years, tests have been conducted on drilling mud wastes from a Superfund site, refinery slop oil, and petroleum sludge to determine whether the process was suitable for treating soils and sludges contaminated with organic compounds. Treatment of the drilling mud wastes was evaluated under the US EPA's SITE Program in 1991 using a pilot-scale system (US EPA 1992a). The pilot-scale system being used to test such materials, located in New Jersey at the Dehydro-Tech corporate headquarters, has a capacity of approximately 45 kg/hr (100 lb/hr) (US EPA 1992a). A trailer-mounted demonstration unit is available for on-site testing (Hazardous Waste Consultant 1991).

Over 85 Carver-Greenfield plants have been licensed worldwide (NETAC 1991); 53 of the facilities are designed to dry and deoil slaughterhouse wastes (US EPA 1992a). Commercial-scale units that have been recently installed include the following treatment processes (NETAC 1991):

- petrochemical activated sludge with 15% solids (installed in 1980 in Italy, with a 6 dry tonne/day (6.5 dry ton/day) capacity);

- biosludge and wool scouring waste with 2% solids (1985, Virginia, 18 tonne/day (20 ton/day));
- dye wastes with 14.3% solids (1986, Soviet Union, 43 tonne/day (48 ton/day));
- alum sludge with 3% solids (1987, California, 15 tonne/day (17 ton/day));
- undigested municipal sludge with 20% solids (1989, Japan, 45 tonne/day (51 ton/day));
- digested municipal sludge with 7% solids (1990, New Jersey, 45 tonne/day (51 ton/day)); and
- oily industrial sludge with 10% solids (1991, Italy, 3.8 tonne/day (4 ton/day)).

When applied in drying wastes, the process can be used to produce fuel, as well as to recover materials such as lanolin (Holcombe and Kollitides 1991).

### 3.10.3 Design Data

Design data for the Carver-Greenfield Process were gathered from several sources, but much of the information was proprietary, and therefore, not available for review. In an economic analysis based on operating data from existing plants, the US EPA estimated that 1.7 billion Joule would be used per tonne of waste (1.47 MM Btu/ton) in treating a waste consisting of 52% solids, 17% indigenous oil, and 31% water by weight. Nitrogen consumption (for deoiling) was estimated at 45.9 standard $m^3$/hr (1,620 standard $ft^3$/hr). The process requires 480-volt, three-phase electric power (US EPA 1992a). Typical operating parameters included the following: energy consumption of 700 J/g to 1,200 J/g (300 to 500 Btu/lb) of water evaporated; operating temperatures of 43° to 177°C (109° to 350°F), and operating pressures of 13.8 to 34.5 kPa (2 to 5 psia) (NETAC 1991). According to DTC, operating pressures may range from 10.3 to 103.5 kPa (1.5 to 15 psia) (Trowbridge 1992).

The pilot-scale unit that has been used in recent treatability testing has a capacity of 45 kg/hr (100 lb/hr) (US EPA 1992a). The system has a nominal capacity of about 45 kg/hr (100 lb/hr) of water evaporation and 180 kg/hr (400 lb/hr)of solvent evaporation (Trowbridge, Holcombe, and Kollitides

1991). The centrifuge and desolventizer process solids at a rate of 13.6 kg/hr (30 lb/hr) (Trowbridge, Holcombe, and Kollitides 1991). In another economic analysis, the US EPA postulated a commercial system could operate at a capacity of 0.92 $m^3$/hr (1.3 $yd^3$/hr), (US EPA 1992d, 36).

Existing operating commercial units worldwide are used to convert wastes to fuels (Holcombe and Kollitides 1991). Feed solids for these units generally ranged from 2 to 20% (although one unit treated 40% solids and another treated 10 to 50% solids) and treatment capacity ranged from 0.3 to 382 dry tonne/day (0.33 to 420 dry ton/day) (Holcombe and Kollitides 1991).

The commonly-used solvent Isopar-L, has a boiling point of 204°C (400°F) (US EPA 1992a). Use of other solvents has been reported. For example, iso-octanol was determined to be effective in extracting bitumen from peat (Holcombe and Kollitides 1991). The solvent-to-feed ratio varies with solids content and treatability goals. In one study, a 16:1 solvent-to-solids ratio was used (see Case Study #2, Appendix C), which is higher than the 5:1 to 10:1 solvent-to-waste solids ratio guideline in US EPA documents (US EPA 1992a). In the US EPA SITE demonstration, a 10:1 ratio of solvent-to-feedstock solids was intended (US EPA 1992a).

According to DTC, pilot-scale systems operate at temperatures of 65° to 93°C (150° to 200°F) in initial evaporation/extraction stages, increasing to 110° to 135°C (230° to 275°F) to evaporate water. At 180°C (360°F), carrier solvent removal from solids in the desolventizer occurred (Trowbridge, Holcombe, and Kollitides 1991).

See Appendix C for case studies which provide indications of temperatures and other operating parameters. Case Study #2 reports on the treatment of refinery slop soil at 80° to 110°C (180° to 230°F) with a vacuum of 280 mm of Hg (US EPA 1992d). The resulting solid samples were deoiled at 149°C (300°F) and 737 mm of Hg vacuum. Solvent was evaporated from the solvent and indigenous oil mixture at about 120°C (250°F) and 600 mm of Hg.

In case Study #3, two stages of evaporation for treatability testing of a petroleum sludge were simulated. The first stage operated around 66°C (151°F) with a vacuum of 585 mm Hg. The second stage operated at 93°C (200°F) with a vacuum of 280 to 305 mm Hg (US EPA 1992a). Following evaporation, the solids were slurried with solvent, filtered, and then deoiled in a vacuum oven at 120°C (250°F) and 710 mm of Hg (US EPA 1992a).

The commercial plant in Case Study #5, having five evaporation stages treating wool scouring wastes, was designed with operating temperatures of 49°, 60°, 71°, 82°, and 127°C (120°, 140°, 160°, 180°, and 260°F) (US EPA 1992a).

### 3.10.4 Pre- and Posttreatment

Materials having a dimension greater than 6 mm (0.25 in.) cannot be treated by the Carver-Greenfield Process, but may be ground to a treatable size (US EPA 1992a).

In the US EPA SITE demonstration, the wastes were slurried with carrier solvent in the first processing step. The developer, however, has described commercial plants in which several evaporative stages involving slurrying with the carrier solvent were conducted before the evaporative and extractive stage (see Case Study #5, Appendix C). As a result, evaporation or dewatering may be considered a pretreatment step for some wastes.

The Carver-Greenfield Process produces the following residues:

- concentrated indigenous oil and organics. According to the developer, this material can either be refined and reused or burned (either for destruction or steam production) (US EPA 1992a);
- water, substantially free of solids and oils. Although the water is described in reports as "substantially free" of solids and oils, it will generally require further treatment. This can usually be accomplished, however, by a wastewater treatment plant (US EPA 1992a); and
- clean, dry solids. The solids may generally be used in landfills, unless this is precluded by regulations or the presence of inorganic compounds. If inorganic compounds such as metals remain, the waste may be subjected to further treatment such as chemical fixation (US EPA 1992d). The removal of petroleum contaminants can greatly improve the effectiveness of chemical fixation (Holcombe, Cataldo, and Ahmad 1990). If the waste feed is classified as hazardous, the treated solids must be characterized prior to disposal (US EPA 1992a).

In addition to these residues, vent gases may require treatment, for example, by granular activated carbon. The concentration of residual organics on the solids will depend on such factors as the influent feed, the type of extracting solvent, and the number of extraction stages (Holcombe, Trowbridge, and Rawlinson 1991).

The developer has described an example materials balance for the entire system (based on tonne per day). For a 91-tonne feed consisting of 10% solids, 40% oil, and 50% water, the materials generated or remaining after each processing step were described as follows (Holcombe, Trowbridge, and Rawlinson 1991, 4):

- 49.5% removed from evaporation stage, consisting of 45 tonne (50 ton) water;
- 36.5% removed as centrifuged concentrate, consisting of 33 tonne (36 ton) oil and 0.3 tonne (0.36 ton) solids;
- 14.0% removed as centrifuged solids, consisting of 8.8 tonne (9.7 ton) solids, 3.5 tonne (3.8 ton) soil, and 0.5 tonne (0.6 ton) water.

The centrifuged solids are desolventized, and makeup solvent (0.3 tonne) is added, resulting in oil-free solids (8.8 tonne solids, 0.1 tonne solvent, and 0.5 tonne water) and oil (3.5 tonne oil and 0.2 tonne solvent) (Holcombe, Trowbridge, and Rawlinson 1991); and

According to DTC, in some instances, the solvent can be a partial cut of the indigenous oil in the feed, which can be recycled within the system (Trowbridge 1992).

### 3.10.5 Operational Considerations

The Isopar-L carrier solvent used in demonstrations of the Carver-Greenfield Process is a food-grade isoparaffinic oil with a high boiling point (204°C (400°F)) and low toxicity. Therefore, special handling is not required (US EPA 1992a). Other solvents with boiling points between 180° and 240°C (360° and 460°F) may be used; however, only iso-octanol has been specifically identified (Trowbridge 1992). The extraction efficiency can be expected to be increased by using particular solvents and additives for particular feed wastes (US EPA 1992a).

Treated solids and contaminated-carrier solvent may be stored in 210 L (55 gal) drums or other containers. Water may also be stored or, alternatively, routed to a POTW. In the SITE demonstration, the final solids and recovered water were basically free of toxic contaminants and therefore, no special handling was required. The residual solvent, which contained extracted organics, required handling as a hazardous waste (US EPA 1992a).

### 3.10.6 Environmental Impacts

The Carver-Greenfield Process is designed to treat wastes on site. The treatment of resulting wastewater and disposal of solids may take place on site or off site (US EPA 1992a). If the original feed is a listed waste, solids residues will be considered listed wastes. If the solids are not listed wastes, they may require testing to determine whether they are RCRA characteristic hazardous wastes. Land disposal regulations must be considered in disposing of these waste streams. Solvent may either be recycled or disposed, and air emissions must be monitored (US EPA 1992a).

Potential environmental impacts include air emissions, dust releases, and transportation issues (US EPA 1992d). The potential for air emissions is mitigated because the process is closed. The developer observes that concerns about air pollution, personnel safety, and odors are minimized because the process is "completely enclosed and operates under slight negative pressure" (Holcombe, Cataldo, and Ahmad 1990). Fugitive dust emissions can generally be mitigated (US EPA 1992a), although they require treatment of the final product with additives (Lau 1991). In sum, the primary environmental impacts occur during removal of recovered indigenous oil and contaminated solvent from the site (US EPA 1992a).

## 3.11 *Extraksol Process*

### 3.11.1 Description

The Extraksol Process is carried out by a transportable system developed by the Sanivan Group of Anjou, Quebec, Canada. The process uses proprietary solvents for batch extraction of contaminated soils. Organic contaminants are extracted in three phases: washing, drying, and solvent regeneration (Mourato and Paquin 1990). See figure 3.5 (on page 3.36) for a simplified process schematic.

**Figure 3.5**
Extraksol™ Process Simplified Process Schematic

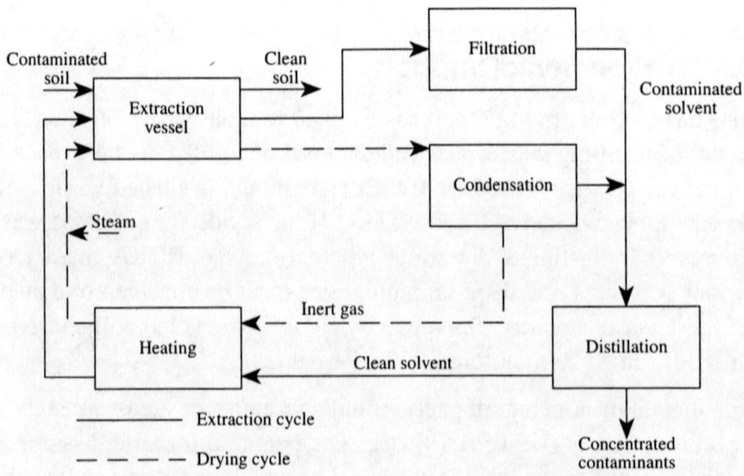


The extraction vessel is a mix tank without an internal agitator. The extractor is loaded with solids, sealed, purged with an inert gas, and filled with solvent. The extraction vessel is then rotated. Absence of an internal mixing system, according to the developer, enables the system to process large nonporous solids. Agitation, combined with the dewatering effect of hydrophilic solvents, will fracture porous solids, reducing their size. After each extraction, the vessel's rotation is stopped, and the solids settle to the bottom. The contaminated solvent is decanted and pumped to a still for solvent recovery, completing a wash cycle. Multiple wash cycles (extraction phases) may be required to meet cleanup requirements for certain contaminants. The number of wash cycles and the appropriate solvent, or blend of solvents, to be used in treating a particular waste stream is determined by bench and/or pilot tests.

The solids drying cycle begins with the draining of the solvent from the extraction vessel through a geotextile filter. Plugging of this filter has not been identified as a problem. This may be due to the fact that the Developer limits application of this technology to solids with no more than a 30%

clay fraction. The solids are heated by introducing hot nitrogen gas and steam into the extractor. The extraction vessel is then rotated. The heated gas strips the solvent from the solids, leaving a dry product. The exact amount of steam to be introduced into the gas stream depends upon the soil and contaminant characteristics. While the additional heat provided by the steam aids volatilization, excess moisture complicates solvent removal and solids drying. Gas is withdrawn from the vessel by vacuum and discharged to a condenser. Condensed solvent and water are discharged to a contaminated solvent tank, while the gas stream flows into a gas holding tank where it remains ready for reuse. At the completion of the drying cycle, the extraction vessel is opened, and the dry solids are discharged to a transport vehicle or conveyor system. The Extraksol unit, operating in the drying cycle only, can be used as a thermal desorber to remove volatile contaminants from solids.

The solvent regeneration cycle begins with the pumping of the contaminated solvent from its holding tank to the distillation unit. The mixture is heated to separate low-boiling point solvents from the high-boiling point components. The low-boiling point components are volatilized, condensed, and collected for reuse as clean solvent. The remaining liquid components flow under gravity through a settling compartment within the distillation unit. Liquid decanted from this vessel is discharged as wastewater and may require further treatment. Still bottoms contain the concentrated organic fraction. As this fraction accumulates, it is pumped to a residue holding tank. It will require further treatment, such as incineration or dehalogenation, depending upon its chemical makeup.

### 3.11.2 Status of Development

Bench-scale treatability tests are conducted to determine whether the Extraksol Process is applicable to a particular waste stream (Mourato and Paquin 1990). The bench-scale test, conducted in a 0.07 $m^3$ (2.5 $ft^3$) rotary mixer, merely simulates the washing cycle. Solids are weighed and loaded into the mixer, the vessel is purged with nitrogen, and a measured volume of solvent is added. The vessel is then rotated in a movement similar to that of the full-scale unit. The mixing time is equivalent to a wash cycle for the full-scale unit. At the completion of a wash cycle, the solvent is sampled and drained from the vessel. A fresh volume of solvent is added, and a second wash cycle is initiated. Any number of wash cycles may be used, although the Sanivan Group normally limits the number of extraction cycles to nine. The cost of full-scale operations using more than nine wash cycles may be prohibitive (Mourato and Paquin 1990).

In addition to the washing tests, the Sanivan Group normally performs material behavior tests consisting of a soil-solvent reactivity test and a filtration test. The soil-solvent reactivity test consists of immersing small amounts of the test solids in a number of different solvents and observing the results. If an emulsion forms, it is filtered through geotextile to determine whether it will restrict solvent flow.

In the filtration test, performed in the rotary mixer, a geotextile filter is placed over the mixer's drain. At the completion of a wash cycle, the solvent is drained through this filter. The time required for the solvent to completely drain from the mixer is noted, along with the volume of solvent and any obstructions and solids particles washing through the filter. A significant amount of solids washing through the filter indicates that the process is not suited to treat the matrix. The Sanivan Group has found that the Extraksol Process is ineffective in treating soils with a clay content >40% and a water content >30%.

The Sanivan Group has constructed two full-scale mobile systems. The 0.9 tonne/hr (1 ton/hr) system is operated with manual controls and is housed on a single 2.4 m by 11 m (8 ft by 36 ft) trailer, may be used in pilot testing, as well as in full-scale treatment. The advantage of using this system as a pilot plant is that material-handling problems are easily identifiable at this scale. Its use is restricted to on-site operations. Support equipment is skid mounted and can be transported on two trailers. The system has been used to treat PCB-contaminated soils; remove oil and grease from refinery sludges, porous gravels, and Fuller's earth; and remove PCPs from porous gravel and activated carbon. Descriptions of these operations have not been published in the literature.

The 5.5 to 9 tonne/hr (6.2 to 10.1 ton/hr) system, constructed in 1991 but not yet used, has an operating unit similar to that of the 0.9 tonne/hr system. Twin extraction vessels are used in this larger, fully automated system. The twin extractors are similar in appearance and operation to cement mixers. Use of twin-batch extractors enables the loading of one extractor while the second is going through its washing and drying cycles. The extractors share a common solvent recovery system (Paquin 1992).

### 3.11.3 Design Data

The Extraksol Process is designed to use several proprietary solvents, individually or blended. This flexibility allows selection of solvents that have a high affinity for particular contaminants. Unit operations consist of:

- washer/dryer extraction;
- nitrogen gas blanketing;
- geomembrane filtration;
- steam generation;
- condensation (refrigerated); and
- solvent distillation.

The number of wash cycles and the wash cycle time may be varied to more effectively remove particular contaminants from the solids. The developer's data suggest that the efficiency in removing PCBs from clay-bearing soils has not exceeded 97.6% (Mourato and Paquin 1990). This is consistent with the developer's suggestion that the process should not be used to treat soils with a clay content >40%. When PCB-contaminated sands were treated, however, only a marginal improvement in removal efficiency (98.6%) was observed (Mourato and Paquin 1990).

### 3.11.4 Pre- and Posttreatment

The process was designed to accommodate large nonporous solids, such as rocks and stones, but porous material must be crushed and/or screened to a particle size of less than 5.1 cm (2 in.) in diameter (US EPA 1992c). Therefore, even though the extraction vessel can accommodate large solids, screening and crushing operations required for particular waste streams, may limit the usefulness of such a feature. In addition, dewatering may be necessary before treating certain waste streams because the process is moisture sensitive. Waste streams with a moisture content >30% can cause reduced removal efficiency, extended time required for solvent regeneration, and increased volume of wastewater (US EPA 1991c).

Posttreatment concerns are essentially the same as those for the B.E.S.T. Process, described in Subsection 3.8.4. The Sanivan Group has stated, however, that chelating agents may be used with the solvent during the wash cycles to mobilize lead (Mourato and Paquin 1990). Chelated lead would be concentrated in the wastewater fraction requiring additional treatment before disposal.

### 3.11.5 Operational Considerations

The 0.9-tonne/hr (1 ton/hr) system is relatively compact, occupying an area less than 28 m$^2$ (300 ft$^2$), exclusive of support equipment. As was previously observed with the B.E.S.T. process, the restricted access zone standard (NFPA 1990) significantly expands the space requirements even for such a compact system. Because of regulations which limit movement of hazardous wastes between Canada and the United States, conducting treatability tests for solid samples which originate in the U.S. may pose some problems. Treatability testing using this system may require obtaining an independent test facility within the U.S. No other unique operating considerations have been identified.

### 3.11.6 Environmental Impacts

To minimize potential air emissions, process gases are vented through an activated carbon filter before being released to the atmosphere. Process wastewaters may contain low concentrations of organic and inorganic contaminants and therefore, may require additional treatment before being discharged.

## 3.12 Low-Energy Extraction Process

### 3.12.1 Description

The following description of LEEP® is based primarily on information provided by ART International, Inc.. The process uses organic solvents to extract organic contaminants from soils, sediments, and sludges. Contaminants are leached from the solid matrix through use of a hydrophilic (water miscible) leaching solvent and then concentrated using either a hydrophobic (water immiscible) stripping solvent or distillation. ART International, Inc. has developed two types of plant processes. The "LEEP-Tar-plant" is targeted for coal tars and related compounds, while the "LEEP-PCB-plant" is intended to treat PCBs and related compounds. The two plants have similar mechanical configurations, but differ in solvent usage and recovery. Figure 3.6 presents simplified process flow diagrams for these two applications.

The process consists of the following steps:

1. Separation and Screening of Contaminated Materials. If feed materials contain free water (moisture content exceeding 50 to 70%), the material is first separated by filtration or centrifugation into a solid and a liquid fraction. For the pilot-scale plant,

**Figure 3.6**

LEEP® Process Configurations

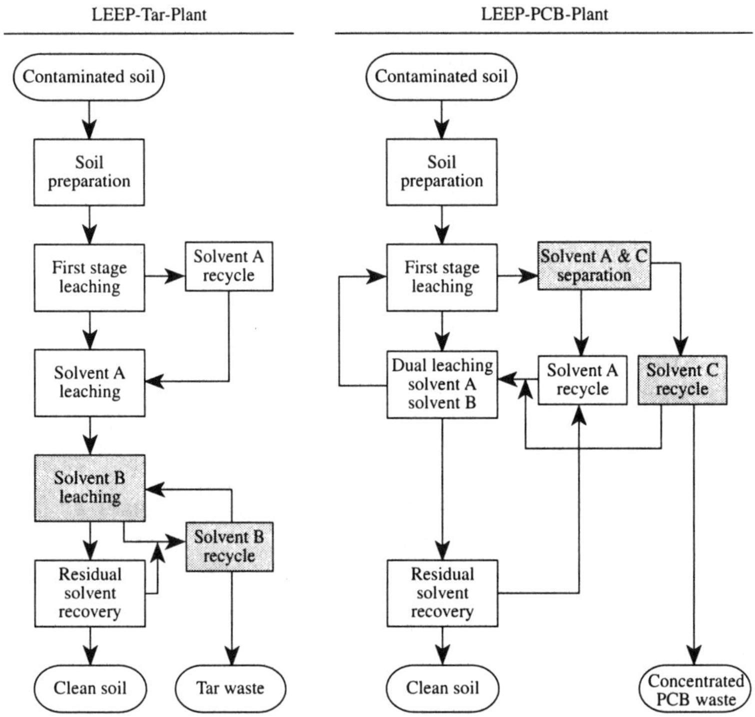


The shaded boxes indicate unit operations which would be substituted in the conversion of a LEEP-Tar-Plant to a LEEP-PCB-Plant

LEEP is a registered trademark of ART International, Inc.

materials must be crushed to less than 6 mm (0.25 in.) size prior to leaching. A commercial (full-size) plant which has been designed, but not constructed, is designed as a closed loop system capable of handling particle sizes up to 203 mm (8 in.) in diameter. Solids are fed through a grizzly/vibrating screen which rejects oversized (>203 mm (>8 in.)) material. The full-size plant includes equipment to crush the solids to 13 to 25 mm (0.5 to 1 in.) size before the materials are fed to the leaching operation.

2a. Treatment of Solid Fraction. Leaching is performed at atmospheric pressure in a continuous solid/liquid countercurrent contactor. The leaching solvent in the PCB-plant is acetone, and in the Tar-plant acetone and a proprietary hydrophobic solvent is used. A number of additional leaching solvents have been tested in bench-scale studies, but are not currently used in the LEEP® process. The level of treatment is adjusted through modifying the flow rate (and therefore the contact time), the revolutions per minute, and/or the liquid to solid ratio. Products of the leaching unit operation consist of clean solids (containing residual clean leaching solvent) and contaminated leaching solvents (containing water, leaching solvent, and the removed contaminants).

2b. Treatment of Liquid Fraction. Any free water is treated through a carbon adsorption water treatment system. The water may be separately discharged or recombined with the soil following completion of soil treatment.

3a. Treatment of Contaminated Leaching Solvents by Liquid-Liquid Extraction or Distillation. For the PCB-plant, the contaminated leaching solvent (acetone) is put in contact with the hydrophobic stripping solvent (kerosene) in order to reduce the volume of contaminated material. Products of the PCB-plant liquid-liquid extraction operation consist of a cleaned leaching solvent and water mixture, containing only trace amounts of contaminants, and a small volume of contaminated stripping solvent (with the majority of the contaminants). Any water leached from the contaminated solids is separated from the acetone by distillation and treated by carbon adsorption in the water treatment system. For the Tar-plant, the contaminated leaching solvents (acetone and proprietary solvent) are separated from water and the contaminants by distillation. Recovered water is treated through carbon adsorption.

3b. Residual Solvent Recovery (Processing of Cleaned Soils). In order to remove residual solvent from the cleaned soil, the solids are treated in a continuous dryer. Evaporated solvent vapors are then condensed, collected, and recycled to the solvent feed of the leaching unit. The soil is then recombined with the treated water and the moist soil is discharged from the process.

4. Stripping Solvent Destruction and Coal Tar Recovery. For the PCB-plant, the contaminant-laden kerosene is collected for off-site disposal. For the Tar-plant, the refined tar from the distillation may be collected for potential commercial use.

### 3.12.2 Status of Development

The LEEP® was developed in 1987 by a New York University research team in the course of a study funded by the US EPA (Steiner and Rugg 1992; Hall, Sandrin, and McBride 1990). The purpose of the study was to develop a low-energy, cost-effective process for removing PCBs from contaminated soils, sediments, and sludges.

The process has been under commercial development by ART International, Inc. since June of 1988 (Rugg 1992). It was accepted by the SITE Program as an emerging technology in June of 1989. By May, 1992, one Canadian and three U.S. patents had been issued and other patent applications were pending (Steiner and Rugg 1992).

Bench-scale treatability studies have been conducted (see Appendix D) and a pilot plant with a nominal throughput of 91 kg/hr (200 lb/hr) was completed in February of 1992. Pilot-plant studies are being conducted under a RCRA permit and a Toxic Substances Control Act (TSCA) permit (US EPA Region II) at the ART International, Inc. facility in New Jersey (Steiner and Rugg 1992). Based on bench-scale and pilot-scale test results, design criteria were prepared for a commercial-size LEEP-Tar-plant. The developer anticipates a plant to be commercially operational by early 1996.

### 3.12.3 Design Data

Bench-scale treatability studies to develop full-scale system design parameters, using 3 to 4 kg of sample, are conducted by ART International,

Inc. The bench-scale studies include three leaching operation tests, one liquid-liquid extraction test, and physical-chemical characterization of the contaminated material (see Appendix D). The exact volume of material required varies with the number of contaminant analyses and the moisture content (Steiner and Rugg 1992).

A variety of leaching solvents have been tested during development of LEEP®, including acetone, n-butylamine, diethylamine, and methylene chloride (Steiner and Rugg 1992). The current leaching solvents which are used by LEEP® include acetone (in the PCB-plant) and acetone mixed with a proprietary solvent (in the Tar-plant). In addition to the leaching solvents, contaminants are separated from the leachate with liquid-liquid extraction using kerosene as a stripping solvent.

As observed in Subsection 3.12.2 a pilot plant which operates at a nominal throughput of 91 kg/hr (200 lb/hr) is available for process testing. Pilot-plant feasibility studies are necessary to determine site-specific throughput rates. The rate of treatment will vary with the matrix type, contaminant levels, and treatment requirements.

ART International, Inc. has developed design criteria for a commercial size LEEP-Tar-plant, and plans additional pilot plant tests with PCB-contaminated soils in order to determine the design criteria for a LEEP PCB-plant. The design capacity of a full-scale mobile Tar-plant process is 11.7 tonne/hr (13 ton/hr (dry)).

The planned commercial size plant is designed to operate at atmospheric pressure and 54°C (130°F). Unit operations are blanketed with nitrogen. According to the developer, the main vent stream is designed to be treated with refrigeration and carbon adsorbers to maintain a near zero % hydrocarbon emission rate. Utility and other requirements are a function of plant size, but are anticipated to be as follows:

- normal battery limit envelope of 1/2 acre;
- 500 to 1,000 hp;
- 8 to 12 million BTU heat load;
- cooling tower make-up of 10 to 20 gpm; and
- 10 to 20 SCFH nitrogen (depending on start-up and shut-down)

According to the developer, LEEP® can be applied to soil moistures of 25 to 35% without reducing throughput rates, and coal tar levels of at least

5% without on-site blending. Higher moisture and tar content soils may also be processed by the unit, although treatment efficiencies will be affected. The LEEP® process is not adversely affected by the presence of cyanide or metals, although the process is not designed to remove these constituents.

### 3.12.4 Pre- and Posttreatment

According to ART International, Inc., the commercial unit has been designed with a front-end preparation section which can handle materials up to 203 mm (8 in.) in diameter. The soil will then be crushed to 13 to 25 mm (0.5 to 1 in.) in size prior to the leaching process. It should be noted that the currently available pilot-scale process is limited to treating finer soils, and cannot process materials exceeding 6 mm (0.25 in.). However, larger materials can be crushed for pilot-scale testing. Due to economic considerations, materials with free water (moisture content exceeding 50 to 70%) are pretreated through centrifugation or filtration.

The residues of LEEP® are handled as follows:

- Clean Solids. The solids generated by the LEEP® plant consist of a clean, moist, solid matrix which may be used as backfill.
- Treated Water. All of the water used in the system is eventually treated by an adsorption unit. This includes the aqueous fraction, which is initially separated from the solid matrix, and water recovered from the water and leaching solvent mixture (after evaporation of the mixture to recover the solvent). Water treated by the LEEP® adsorption unit may require further treatment steps to ensure that all contaminants of concern are removed. According to the developer, additional water treatment unit operations may be incorporated into the LEEP® system.
- Contaminated Stripping Solvent. The contaminants are concentrated in the stripping solvent. The contaminated stripping solvent may be collected for off-site treatment (incineration) or in some cases for reuse.

The leaching solvent is recycled. No information was available regarding potential contaminant releases to the air during processing of the contaminated soils. According to ART International, Inc., however, the system is designed as a closed system which is blanketed with nitrogen. This blanket carries residual leaching solvent through condensers and a carbon adsorption system prior to discharge to the atmosphere. The developer, therefore, does not expect contaminants to escape. It should be noted that treatability testing should be conducted on sample soils in order to identify residue problems specific to a site.

### 3.12.5 Operational Considerations

As with other innovative technologies, bench- and pilot-scale studies should be conducted to determine the potential effectiveness of the process in treating wastes at a particular site. Effectiveness of the LEEP® depends on the particular matrix, as well as the type and level of contaminants present. The rate of treatment (throughput) is affected by both the contaminant-leaching rate and the particle-settling rate. The throughput, in turn, affects costs.

Therefore, the key information required is the characterization of the matrix (percent moisture, particle size), type and level of contaminant (site-specific), and required treatment levels.

The reagents used in this process (acetone and kerosene) are highly flammable. Therefore the design and operation of the system must include measures to minimize the potential for fires.

### 3.12.6 Environmental Impacts

Detailed information concerning effects of the process on air, water, and land was not available. The developer claims that no contaminant releases are expected because the system is closed. Bench-scale test data indicate the process will result in clean soil and water discharges.

A potential environmental impact may arise from the initial material preparation and crushing activities. However, the developer indicates that this front-end treatment will be conducted under a slight negative pressure in order to control dust and fugitive emissions to meet regulatory requirements.

The final stripping steps for both the soil (solvent evaporation) and water (adsorption unit) must be carefully monitored to assure that the soil and water residues do not release leaching or stripping solvents at the site. As described in Case Study #1 (Appendix D), some solvent schemes may reduce specific contaminants, but elevate levels of other compounds (semivolatile remnants and other TICs). Consequently, treatability testing should be conducted. When the leaching solvent is successfully removed from soil residues, it is not expected to have significant environmental impact because it is recycled within the system. The stripping solvent (which contains the removed contaminant) is collected for off-site destruction (incineration). Thus, if performed with sufficient care, this stage poses no on-site environmental threat. There is, however, potential for off-site impact in the form of air emissions resulting from incineration of the removed contaminants.

# 3.13 NuKEM Development Process

## 3.13.1 Description

NKD, a division of American NuKEM, has under development two waste treatment processes employing solvent extraction. One process, targeted primarily at remediation of contaminated soil, employs mixer/settlers for the extraction operation (Massey and Darian 1989). The second process, targeted at wastewaters and sludges from petroleum refineries, employs a continuous extraction column (Massey and Darian 1990).

Figure 3.7 (on page 3.48) provides a schematic diagram for the contaminanted soils process. In the initial step, contaminated soil is fed to a mixer where it is combined with a proprietary solvent and a proprietary chemical reagent. Depending on the type of soil, the solvent-to-soil ratio varies from a typical ratio of 1:1 to a maximum of 2:1 by weight. The slurried soil leaving the mixer is fed to a countercurrent extraction unit employing mixer/settlers. In this unit, the soil is countercurrently combined with solvent, and the organic contaminant is progressively removed as it passes through the extraction stages. The number of extraction stages employed is determined by the degree of decontamination desired. Three to five stages of extraction are normally adequate to reach target levels.

The treated soil leaving the last stage of the countercurrent extraction unit is fed to a solvent dryer in which the solvent is volatilized from the soil, condensed, and recycled, either to the initial feed treatment step or to the last stage of the extraction unit. The system is intended to be transport-

**Figure 3.7**
NKD Process Schematic — Contaminated Soils

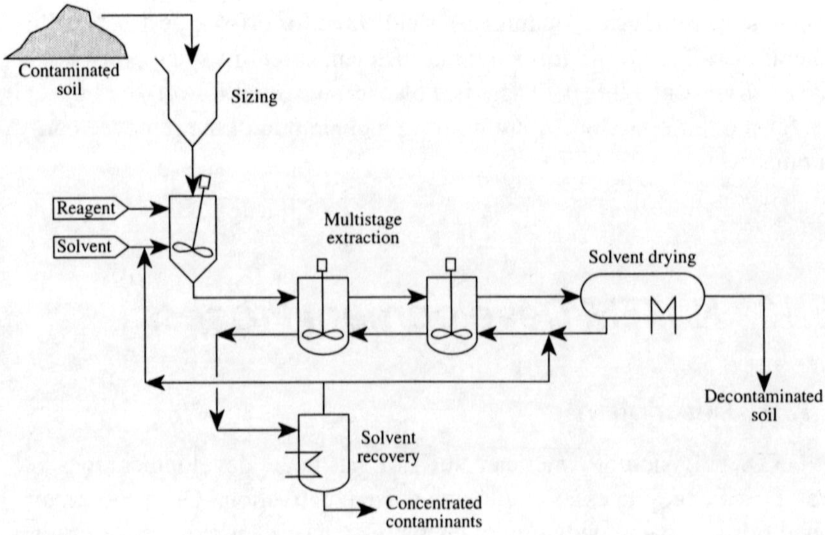


able and would normally be operated on site. Accordingly, the treated soil leaving the solvent dryer would normally be stored in a product storage area, checked for quality, and returned later to the excavation from which removed.

The contaminated solvent (extract) is withdrawn from the first mixer/settler and is processed in a distillation unit in which the contaminants are separated from the solvent. The recovered solvent is recycled to the initial feed treatment step. The residual contaminants can be either destroyed on site or, more typically, put in drums in a concentrated form and shipped to an incinerator for destruction (Massey and Darian 1989).

A variation of the process is under development for treatment of refinery waste streams having RCRA waste codes K048, K049, K050, K051, and K052. This process is designed to directly treat raw refinery oily wastes, no pretreatment being required. The vast majority of refinery oily wastes exist as bulk water streams containing small amounts of oil and solid. Under most processes, these streams are dewatered before treatment, an unneeded operation under the NKD Process. In this variation (see figure 3.8), extraction is carried out in a multistage column extractor. Raw sludge (70% water, 15% solids, 15% oil) is combined with a small amount of refinery-based solvent and an unspecified amount of the proprietary reagent. This mixture is fed at the top of a multistage, liquid-liquid countercurrent extraction column. For a 38 L/min (350 bbl/day) plant, this column would be 76.2 cm (30 in.) in diameter and 10.7 m (35 ft) in height. Solvent is fed at the bottom of the extraction column at a nominal flow rate of 75 L/min (20 gal/min). Solids and water flow down the column, while solvent flows from the bottom to the top, extracting oil from the refinery waste as it rises. The oil-laden solvent (extract) exits the top of the column, while oil-free solids and water exit the bottom. The solids and water (raffinate) from the bottom

**Figure 3.8**
NKD Process Schematic — Oily Waste

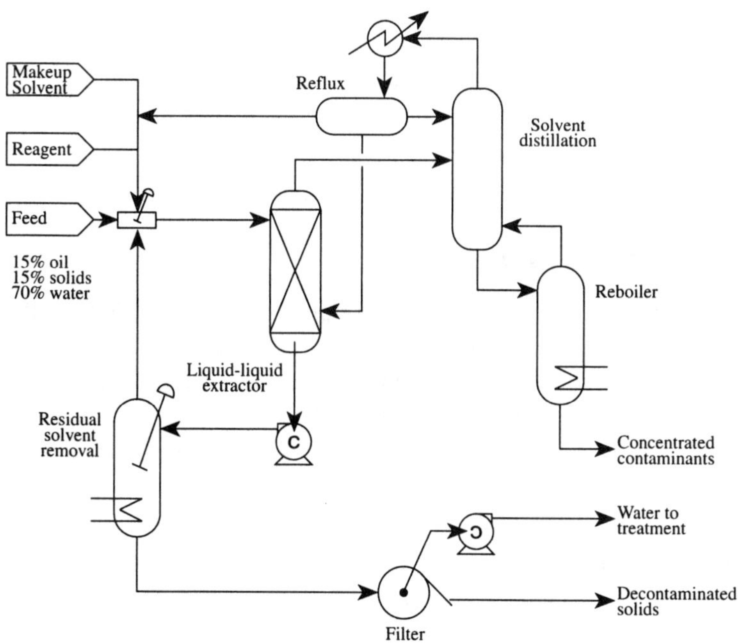

of the extraction column are pumped to a solvent stripper (residual solvent removal operation in figure 3.8 on page 3.49) in which residual solvent is stripped from the mixture. The resulting slurry of solvent-free solids and water is then pumped to a filter. Water from the filtration step is sent to the refinery's wastewater treatment system, and the filter cake is accumulated in a roll-off bin for shipment to a disposal site for stabilization, if necessary, and then to landfills (Massey and Darian 1990).

The extract is fed to a fractional distillation column for the recovery of the solvent. Reasonable recoveries are achieved at a reflux ratio of about unity. Recovered oil from the reboiler is recycled to the refinery crude oil fractionation unit, coker, or fluidized catalytic cracker.

One option, which greatly simplifies the process, is to return the extract directly to the refinery crude oil fractionation unit. If this is done, the system would not include a solvent distillation unit; the refinery would provide fresh solvent and would receive back the oil-laden extract stream.

NuKEM Development states that in general, solvent extraction processes tend to operate much less efficiently in the presence of water than in its absence. The difficulty increases greatly when such processes are used to treat soils containing >10% water. The water, soil, and solvent tend to form emulsions and either agglomerate or coat the walls of the process vessels. Consequently, the solvents do not adequately contact the soil and do not effectively extract the contaminants. NuKEM Development claims to have solved this problem by adding a small quantity of a proprietary reagent to the soil in the initial process step.

### 3.13.2 Status of Development

As of August 1989, NKD was conducting pilot-scale studies on its soil decontamination process at its Houston facility (Massey and Darian 1989). In October 1990, pilot-scale efforts were underway on the refinery waste version of the technology. A 480 L/day (4 bbl/day) pilot plant was being operated in Houston (Massey and Darian 1990). See Appendix E for a report of the treatability studies' results of the NKD Process.

### 3.13.3 Design Data

Very little design data on the NKD Process are available. Pilot-plant operation of the refinery waste process employed a 7.6 cm (3 in.) diameter, continuous, countercurrent extraction column containing a total of 36 actual extraction stages in an overall column height of 1.2 m (4 ft). The throughput rate was 320 to 640 L/day (2 to 4 bbl/day), but apparently the system was operated at higher rates from time to time. A feed rate of up to about 75 L/hr (690 bbl/day) is reported elsewhere (Massey and Darian 1990).

Based on the pilot-plant study results, NKD has projected the key features of a process to treat the oily wastes from a 16 MM L/day (100,000 bbl/day) oil refinery. Raw API separator sludge and DAF sludge production would be 9,100 tonne/yr (10,000 ton/yr) and slop oil production would be 910 tonne/yr (1,000 ton/yr). The operation would require 160,000 L (42,000 gal) of storage for sludge and 16,000 L (4,200 gal) for slop oil.

Sludge would be drawn from the storage tank at about 38 L/min (10 gal/min) for about four days per week. During the fifth day, slop oil would be processed at rates ranging from 3.5 to 19 L/min (1 to 5 gal/min), depending on the characteristics of the sludge. The remaining days of the week would be devoted to peripheral activities, including recharging the storage tanks, characterizing the next feedstock, and establishing parameters for the next week of operation.

The process flows as depicted in figure 3.8 (on page 3.49) would have the following flowrates when operating on the sludge stream at 38 L/min (10 gal/min). The solvent extraction column would be 0.76 m (2.5 ft)in diameter and 10.8 m (35.4 ft) in height. Solvent would be injected into the base of the column at a nominal flow rate of 76 L/min (20 gal/min).

The distillation column for fractionation of the extract has not been sized, but would operate at a reflux ratio of near unity and would have a heat duty in the 1,582 to 2,637 Joule (1.5 to 2.5 MM Btu/hr) range, with the exact value depending on the degree of heat integration employed. Alternatively, the extract could be returned to the refinery crude oil fractionation unit, simplifying the process (Massey 1990; 1992).

### 3.13.4 Pre- and Posttreatment

Excavated soil is screened to remove bulk rock, tree stumps, and debris of like size. The remaining soil is reduced to <5 cm (2 in.).

The soil remediation process produces clean soil and two by-product streams. One by-product is debris and the other is the concentrated organic waste, including the PCBs. Originally sold as a mixture with oil, PCBs are generally recovered as a similar mixture. This stream usually is a liquid. Most remediation sites also contain other organics, and all organics with vapor pressures less than that of the solvent should appear in the concentrated waste stream. It is possible that very light organic compounds could be present, and that these could build up in the solvent stream, which is recycled within the process. If this happens, it might be necessary to purge some of the solvent from the process (Massey and Darian 1989).

For the refinery waste treatment process with a feed rate of 2,268 kg/hr (5,000 lb/hr) of sludge and 34 kg/hr (175 lb/hr) of solvent, the following streams are expected:

- recovered oil (374 kg/hr (825 lb/hr) of which 340 kg/hr (750 lb/hr) is oil and 34 kg/hr (75 lb/hr) is solvent);
- filtered water (1,361 kg/hr (3,000 lb/hr)); and
- filtered solids (567 kg/hr (1,250 lb/hr), of which 340 kg/hr (750 lb/hr) are solids and 227 kg/hr (500 lb/hr) is water).

Again, it is possible that high-volatility compounds could build up in the solvent stream (Massey and Darian 1990).

### 3.13.5 Operational Considerations

The initial goal of the NKD development effort was to extend the company's capability for treating PCB-contaminated soils and sludges. The technology was to be capable of achieving treated soil PCB levels of less than 2 ppm and be highly mobile rather than merely portable. Other goals were that the process be simple, scalable, and economical in treating small, as well as large, quantities of contaminated soil. At the time the development effort was begun, it was felt that the problem of handling high-moisture content soils was unsolved. NKD believes their process is capable of handling soils with a wide range of naturally-occurring moisture content as if no moisture were present (Massey and Darian 1989).

No unusual health considerations are evident. Exposure of employees to the waste and to the refinery-derived solvent should be minimized. The proprietary reagent used is said to be nontoxic and safe to handle. The solvent is flammable; process equipment should be designed and operating procedures devised accordingly.

### 3.13.6 Environmental Impacts

The process does not appear to have potential for any unusual impact on the environment. The solvent employed must be handled properly and storage tanks must have appropriate venting devices. To prevent contamination of nearby aquifers, the solvent must be thoroughly stripped from the treated soil before it is placed back into the excavation site. With a properly designed process, solvent losses to the atmosphere, soil, or aquifers should not be significant. Proper solvent handling, secondary containment, and other appropriate design features should mitigate impacts on the soil and the groundwater in the vicinity of the operation.

## *3.14 Soil Restoration Unit*

### 3.14.1 Description

Terra-Kleen Corporation developed a Soil Restoration Unit (S.R.U.) that uses up to 14 different organic solvents in treating contaminated solids. The solvents to be used in extracting organic contaminants from a particular waste stream are determined through a series of bench-scale treatability tests. The solvent is selected based upon the solubility characteristics of the contaminant(s) and its phase separation characteristics with respect to the solid matrix (Cash 1991).

The process steps are the same as those of the Extraksol Process: washing, drying, and solvent regeneration. The S.R.U. Process, unlike a batch process, conveys the solids through a specially-designed extraction unit where they are mixed with solvent(s) (see figure 3.9 on page 3.54). Conveyors are used to move solids from the solids feed hopper to the extractor and then on to a solids drying unit. Fresh solvent continuously circulates through the extractor. As the solvent mixes with the solids, organic substances are solubilized and removed. Continuous flushing of the contaminated solvents with fresh solvent causes solute concentrations in the solids to equilibrate at a low level with solute concentrations in the liquid. Since fresh solvent has no solute (organic contamination), removal is limited by diffusion or desorption of the contaminants from the solid particles. Extrac-

**Figure 3.9**
Terra-Kleen Soil Restoration Unit Process Schematic

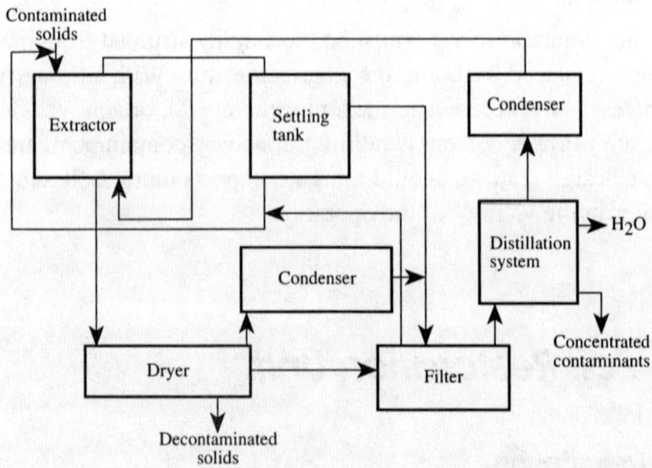


tion time can be adjusted to optimize removal efficiencies and minimize processing time.

The solids drying unit is designed to remove residual solvent and produce a clean, dry product. The drying chamber is an enclosed tube through which a heated air and nitrogen mixture is blown. Solvent vapors vented from the dryer are condensed and drain to the solvent regeneration system. Toward the end of the drying cycle, dried soil particles may become airborne. When this occurs, the air and vapor flow can be diverted through a bag filter located ahead of the condenser. The filtered and condensed gas stream is reheated and reused as feed gas for the dryer. Baghouse dust is discharged as a clean solid.

### 3.14.2 Status of Development

Terra-Kleen Corporation uses a bench-scale S.R.U. for treatability testing. Solids are loaded into an extractor, the extractor is sealed, and solvent is circulated through the extractor. The extraction solvent is sampled at specified intervals and analyzed for contaminant(s) content. When the solvent samples show no detectable concentration of contaminant(s), the solids are sampled and analyzed. Up to 14 different solvents or solvent blends can be used during the test.

In 1988, the Terra-Kleen Corporation built a trailer-mounted pilot-plant system that was designed to remove organic contaminants from soils. This system, used in a Superfund removal action at the Traband site in Oklahoma, served as the prototype for the S.R.U. and has since been dismantled (Cash 1991, 1992).

The Terra-Kleen Corporation has designed and constructed a self-contained mobile S.R.U. capable of treating up to 1.8 tonne (2 ton) of contaminated solids per hour. The mobile unit is housed within two enclosed 2.4 m by 12 m (8 ft by 40 ft) trailers and carries out all process operations — washing, drying, and solvent regeneration. Support equipment consists of solvent holding tanks, solids classifiers, premixing tanks, and an auxiliary distillation system, the need for which depends on the particular site.

A full-scale system was completed in the spring of 1992. It was scheduled to treat approximately 2,700 tonne (3,000 ton) of soils contaminated with PCBs and chlorobenzenes at the Pinette Salvage Yard Superfund Site in Washburn, Maine. During the summer and early autumn of 1992 the S.R.U. was operated at the Pinette Salvage Yard Superfund Site. Soils contaminated with PCBs were treated to achieve a clean up goal of 5 mg/kg. This goal was achieved; however, the design throughput for the unit was never realized. This was due to a series of problems which plagued the operation. Oblong-shaped rocks, which were approximately 5 cm in diameter and 7 to 15 cm in length, passed through the soils screens and broke the conveyor system. Several days of operation were lost while the system was repaired. Fines from the solids did not separate from the solvent during the settling phase. Therefore, the developer had to replace the settling chamber with a second unit which had a greater capacity. A polymer feed system was also added which aided in flocculation and settling of the fines. Finally, the drying system did not remove all of the residual solvent from the treated solids, and landfarming techniques had to be used to biologically degrade the residual solvent before the treated soils could be returned to the site.

Terra-Kleen has examined the operating problems which were experienced at the Pinette Salvage Yard Superfund site and has designed a new solvent extraction system. This system is batch fed, to eliminate material handling problems, and uses a vacuum-extraction system to remove residual solvent from treated solids. A SITE demonstration of this technology was planned for the summer of 1994.

### 3.14.3 Design Data

In addition to using up to 14 different solvents and solvent blends, according to Terra-Kleen Corporation, the S.R.U. can use washing fluids that use surfactants and/or chelating agents for mobilizing inorganic contaminants such as metals (US EPA 1991c). The following are the unit operations:

- solvent/solids premixing;
- continuous solvent flow extraction;
- heated air/nitrogen drying;
- condensation;
- bag filtration;
- solvent settling;
- solvent filtration; and
- distillation.

The specific solvent(s) to be used and the time required for extraction of particular wastes is initially determined through treatability tests. But the regime can be adjusted based upon results observed during operation. A pilot-scale prototype was used in treating PCB-contaminated sandblasting sand at the Traband site in Oklahoma. A cleanup goal of less than 100 ppm PCBs was established for the three sand piles at the site. Analysis of samples from each sand pile revealed that concentrations of PCBs before treatment were 4,600, 3,300, and 88 ppm. Following treatment, the concentrations of PCBs in composite samples were 94, 47, and 4 ppm (Cash 1991).

### 3.14.4 Pre- and Posttreatment

This process is designed to accommodate solids up to 7.6 cm (3 in.) in diameter. Therefore, before a particular waste stream is treated, solids may have to be screened and/or crushed. The size of the solids particles within the extractor will be further reduced by solvent dewatering and by mechanical forces of the material handling equipment. Additionally, waste streams with high-contaminant concentrations (>3,000 ppm of a particular contaminant) may require premixing with the selected solvent (Cash 1992). Premixing effects intimate contact between the solids particles and the solvent, thereby reducing the overall time required for extraction.

The moisture content of a particular waste stream may also affect extraction of organic contaminants. High-moisture content solids (>30%) can dilute hydrophilic solvents, thus reducing their effectiveness in solubilizing organic contaminants. Furthermore, since distillation is employed to separate water from solvent, the higher the initial moisture content, the greater the energy required for separation.

Solids discharged from the S.R.U. are dry, but some residual solvent and organic contamination may remain. Solids meeting cleanup criteria may be returned to the land for reuse. Since beneficial organic matter may also be removed by the extraction process, augmentation of the treated solids with organic substrate may be necessary to promote degradation of residual solvent. Inorganic contaminants are not removed from solids when organic solvents are used for decontamination, and therefore, the solvent-extracted solids may need to be treated by some other technique, such as stabilization/solidification or soil washing (Meckes et al 1992; US EPA 1990c).

However, the Terra-Kleen Corporation has stated that chelating agents may be used with the solvent during the wash cycles to mobilize metallic contaminants. Chelated metals would be concentrated in the wastewater fraction, which would require additional treatment before disposal.

Water produced through this process should be analyzed for the contaminant(s) of concern and residual solvent. Low concentrations of organic contaminants may remain with the water fraction. These contaminant concentrations may be sufficiently low to permit discharge to a POTW. Liquid-phase carbon adsorption may be used to produce an aqueous discharge free of measurable organic contaminants.

Organic contaminants are concentrated in the still bottoms. If chlorinated hydrocarbons, such as PCBs and some pesticides, are among the contaminants of concern, the treatment options are incineration or dehalogenation. If chlorinated hydrocarbons are not among the contaminants of concern, the concentrated waste stream may be recycled.

### 3.14.5 Operational Considerations

The 1.8 tonne/hr (2 ton/hr) system is relatively compact, occupying an area of approximately 93 $m^2$ (1,000 $ft^2$), exclusive of support equipment. Use of an auxiliary distillation unit can add an additional 37 $m^2$ (400 $ft^2$) to the siting requirements (Cash 1991). The solvents used are flammable; therefore, NFPA standards are applicable. No other operational considerations have been identified for this system.

### 3.14.6 Environmental Impacts

Because of similarities in the process design and solvents employed, environmental impacts of the S.R.U. are expected to be essentially the same as for the Extraksol Process (see Subsection 3.11.6).

# 4

# POTENTIAL APPLICATIONS

Solvent/chemical extraction (SCE) processes have been selected by the U.S. Environmental Protection Agency (US EPA) for some Superfund sites contaminated with organics such as polychlorinated biphenyls (PCBs), volatile organic compounds (VOCs), and pentachlorophenol. The SCE processes discussed herein were developed to treat a wide range of organic contaminants in several different matrices. See table 4.1 (on page 4.2) for a summary of the types of contaminants removed in bench-, pilot-, or demonstration-scale testing by the processes reviewed. The development of these processes has typically proceeded from a design addressing a particular problem (PCBs in sediments) to a more general design capable of treating a wide range of contaminants and matrices.

## *4.1 Basic Extractive Sludge Treatment Process*

Numerous Basic Extractive Sludge Treatment (B.E.S.T.) Process bench-scale treatability tests have been conducted on soils, sludges, and sediments contaminated with PCBs, polyaromatic hydrocarbons (PAHs), pesticides, and other semivolatile and volatile organic contaminants. The results show that highest removal efficiencies were achieved in treating solids that had high initial concentrations of organic contaminants. In many cases, however, the treated solids retained a significant amount of the initial contaminant. For example, tests of three harbor sediment samples contaminated with PCBs in concentrations of >20,000 mg/kg resulted in removal efficiencies of >99.8% after three extraction stages, but with residual PCB concentrations in the solids from 27 to 720 mg/kg. On the other hand, treatment of two sediment samples having initial concentrations of PCBs of 83 and 68 mg/kg resulted in removal efficiencies for both samples of 99.6% and re-

**Table 4.1**
Potential Applications
Commercial Solvent/Chemical Extraction Processes

| Contaminant Type | Matrices Tested |
|---|---|
| PCBs | Soils (sands, loams, clays) |
| PAHs | Sediments |
| VOCs | Sludges |
| Semi-VOCs | Slurries |
| Pesticides | Wastewaters<br>Drilling cuttings<br>Petroleum-listed wastes |
| Pentachlorophenols | K044 - K052 wastes* |
| BTX | |
| Dioxins | |
| Diesel fuel | |
| Petroleum Hydrocarbons | |

* K044 - K052 are RCRA waste codes.

sidual PCB concentration in solids of 1.1 and 1.8 mg/kg. All of this points to the need to conduct treatability tests before selecting this technology for site remediation.

The B.E.S.T. Process may also be used in treating sludges from petroleum refineries and petrochemical operations, including Resource Conservation and Recovery Act (RCRA) waste codes K044 through K052, inclusive.

## 4.2 *CF Systems*

The CF Systems's supercritical technology uses liquefied gases as extracting solvents to remove organic contaminants, such as hydrocarbons and oil and grease, from wastewaters, sludges, sediments, and soils. Carbon dioxide is generally used for aqueous solutions, such as process water and wastewater. Light hydrocarbons are recommended for sludges, sediments, and soils. Supercritical technology can be applied to a large variety of organic contaminants, including carbon tetrachloride, chloroform, benzene, naphthalene, gasoline, vinyl acetate, furfural, organic acids, dichloroethane, oil and grease, xylene, toluene, methyl acetate, acetone, alcohols, phenol, aliphatic and aromatic hydrocarbons, and PCBs.

CF Technologies, Inc. is now commercializing supercritical fluid extraction and has described possible applications encompassing solid and semi-solid waste treatment, wastewater treatment, and pollution prevention, including the following:

- removal of PCBs from contaminated soil;
- extraction of benzene, toluene, xylene (BTX) in wastewater from petrochemical storage tanks and oil and water separators;
- separation of pesticides from contaminated soil;
- extraction of oil, lubricants, and PAHs from solids and groundwater at synthetic gas and aircraft or automotive maintenance facilities;
- recovery of certain Comprehensive Environmental Response, Compensation, and Liability Act/National Priorities List (CERCLA/NPL) pollutants, i.e., methyl methacrylate and tetraethyl lead; and
- substitution for chlorofluorocarbons (CFCs) and chlorinated solvents in cleaning and degreasing operations.

## 4.3 Carver-Greenfield Process

The Carver-Greenfield Process can be used to remove oil-soluble organics from soils, sludges, and other wastes, as well as to dry aqueous mixtures (NETAC 1991). As noted in the US EPA Superfund Innovative Technology Evaluation (SITE) Program report (US EPA 1992a), the process can be used to treat wastes contaminated with organics, especially wastes with high-water content.

The developer states that Carver-Greenfield has a new approach for remediating soils, petroleum K-wastes, spent drilling muds, and hazardous sludges containing petroleum-based contaminants, such as fuel oils, PCBs, and polynuclear aromatics (US EPA 1992a). It should be noted, however, that most commercial units are designed to treat high-water content wastes. According to Dehydro-Tech Corporation (DTC), existing units "have processing requirements very similar to oily soil/sludge treatment," and DTC observes that dewatered municipal sewage sludge typically contains over 20% solids and has a high-ash content (US EPA 1992a). But, it should also be noted that most contaminated soils and sludges are expected to have a far higher solid content than that of municipal sewage sludge, and the potential effectiveness of the Carver-Greenfield Process in treating low-water content wastes must be investigated further.

Laboratory-scale tests have been conducted on solids from a site contaminated with diesel fuel and PCBs, as well as on mixed wastes (see Appendix C for case studies). In addition, pilot-scale studies have been conducted on oily drilling mud wastes. Commercial units have treated such materials as municipal and industrial sewage sludges, meat rendering waste, wool scouring wastes, petrochemical sludges, wood pulp wastes, dairy and food products, textile and dye wastes, paper mill sludge, brewery treatment plant sludge, animal manure (for fertilizer production), and pharmaceutical wastes. The soils and sludges process is intended mainly to treat those contaminated with oil-soluble hazardous contaminants, including PCBs, PAHs, and dioxins (US EPA 1991b; US EPA 1989).

Some commercial units incorporate portions of the Carver-Greenfield Process designed for unique applications. For example, oily slops and sludges from refineries may be disposed of by injecting them into coker feed. In one design, this method was modified by dewatering the sludge with the Carver-Greenfield Process, prior to coking it. In such a system, before centrifuging the solids or recovering the oil, the oil-solid slurry direct from the evaporative/extraction stage is sent to the delayed coker. The carrier oil may later be recovered from the coker vapor products (Elliot 1992).

## 4.4 Extraksol Process

The developer states that the Extraksol Process can effectively treat soils that have a maximum clay fraction of 40% and a maximum water content of 30% (Paquin 1992).

The number of wash cycles and the wash cycle time may be varied to effectively remove certain contaminants from solids. The developer's data suggest that efficiencies in removing PCBs from clay-bearing soils has not exceeded 97.6% (Mourato and Paquin 1990). This is consistent with the developer's suggestion that soils with clay contents >40% should not be treated by this process. But, only a marginal improvement in efficiency (98.6%) was noted when PCB-contaminated sands were treated by this process.

Refinery sludges were also treated to remove oil and grease with efficiencies of up to 99%. Similar results were obtained in removing grease and oil from contaminated Fuller's earth. Oily sludges contaminated with PAHs were also treated; removal efficiencies of up to 96% were reported, with a final PAH concentration of 10 ppm. Pentachlorophenol-contaminated gravels and activated carbon were also treated using the 0.9 tonne/hr (1 ton/hr) system, with removal efficiencies of >99.7% in the case of gravel and 89% for carbon (Mourato and Paquin 1990). Other results indicate considerable variability in efficiencies in removing oil and greases and in removing particular contaminants. Therefore, site-specific treatability testing is encouraged before the Extraksol Process is used in site remediation.

## 4.5 Low-Energy Extraction Process

The Low-Energy Extraction Process (LEEP®) is designed to treat coal tars and PCB-contaminated materials. The process may also be used to remove other organic pollutants such as petroleum hydrocarbons, PAHs, pesticides, wood-preserving chlorophenol compounds from contaminated soils, sludges, and sediments (Steiner and Rugg 1991). Materials that have been treated by LEEP® include river and harbor sediments, sandy topsoil, clay subsoil, and foundry sand. As noted by ART International, Inc., the system developer, an advantage of the system is its ability to treat materials having moisture content ranging from a few percent to >90% (Steiner and Rugg 1992).

## 4.6 NuKEM Development Process

For the soil remediation version of the NKD Process, results have been published only on the removal of PCBs from soils and sludges. It is expected, however, that the process will be effective in removing a wide range of volatile organics, semivolatile organics, pesticides and their intermediates, petroleum hydrocarbons, and other organics. The petroleum refinery sludge treatment version has been shown to be effective in treating API separator sludge (US EPA waste code K051), dissolved air flotation (DAF) float (US EPA waste code K048), and slop oil sludge (including US EPA waste code K049). The process is expected to be effective in treating other refinery waste sludges (Massey and Darian 1989, 1990; Massey 1992).

## 4.7 Soil Restoration Unit

No operating data are available for the full-scale Soil Restoration Unit (S.R.U.), and data available on the process are limited. A pilot-scale prototype was used in treating PCB-contaminated sandblasting sand at the Traband site in Oklahoma. A cleanup goal of a concentration of <100 mg/kg PCBs was established for the three sand piles at the site. Analysis of samples from each sand pile revealed that PCB concentrations were 4,600, 3,300, and 88 mg/kg. Following treatment, PCB concentrations had been reduced to 94, 47, and 4 mg/kg in composite samples.

Another test was conducted on 15 $m^3$ (20 $yd^3$) of soil, described as a sandy loam contaminated with PCBs in concentrations up to 200 mg/kg. Following treatment, residual PCB concentrations in the treated soil varied between 2.5 and 4.5 mg/kg. During this test, ambient outside temperatures ranged from -15° to -12°C (-10° to 5°F). The prototype system was not insulated and, therefore, suffered heat loss. Terra-Kleen, the system developer, believes that the temperatures decreased extraction efficiencies significantly.

Soil samples contaminated with diesel fuel in concentrations up to 6,190 mg/kg have also been treated. A removal efficiency of 98% was achieved.

# 5 PROCESS EVALUATION

Solvent/chemical extraction (SCE) is effective in separating a wide range of organic contaminants from soils, sludges, and sediments (see table 4.1 on page 4.2), thereby reducing the volume of contaminants, which may require further treatment. In addition, SCE has been demonstrated to be effective in treating a number of RCRA-listed wastes such as wood treating wastes (K001), slop oil emulsion solids (K049), American Petroleum Institute (API) separator sludge (K050), and tank bottoms (K052). The technology has been applied also in treating drilling muds, coal tar wastes, paint wastes, synthetic rubber process wastes, pesticide and insecticide wastes, and others. It has been selected in six Records of Decision to clean up Superfund sites (US EPA 1992e).

SCE is evaluated under the following parameters:

- levels of removal of contaminants;
- status of development; and
- costs.

## 5.1 Levels of Removal of Contaminants

Concentration factors of up to 10,000:1 have been measured. This represents a significant reduction in the volume of contaminants requiring further treatment.

Removal efficiencies and levels of reduction that can be effected vary with the particular process, the number of extraction stages, the type and concentration of contaminants, and the nature of the medium. Reduction of >98% of polychlorinated biphenyls (PCBs) at levels up to 4,600 ppm and reduction of >95% of polyaromatic hydrocarbons (PAHs) at levels up to 2,900 ppm have been reported by the technology vendors. Overall organics recovery efficiencies of greater than 99% are claimed by one vendor (CF Systems 1992).

Removal efficiencies >90% are generally reported for all organic contaminants with residual levels in many cases <1 ppm. This performance, however, may require a high number of extraction stages (6 to 8), particularly when there are very high initial concentrations.

## 5.2 *Status of Development*

Two systems, CF Systems and Carver-Greenfield, have completed US EPA Superfund Innovative Technology Evaluation (SITE) demonstration programs and are fully documented (US EPA 1989; 1992a). One other system, Best Extractive Sludge Treatment (B.E.S.T.), has undergone a 24-hour US EPA evaluation program and a report has been issued (US EPA 1988). Three systems (B.E.S.T., Extraksol, and Soil Restoration Unit (S.R.U.)) are currently in the US EPA SITE Program (US EPA 1992g; Meckes et al. 1992).

All of the system suppliers are offering commercial systems for a wide variety of applications. Most applications require treatability testing to enable site-specific design parameters to be determined. Treatability tests show considerable variation in results from different sample sources; therefore, a well-planned and executed treatability test program is needed to properly set site-specific design parameters (US EPA 1992b).

## 5.3 *Costs*

Unit cost data were solicited from each process vendor and that received was augmented with other published cost data. See table 5.1 (on page 5.3), which reports unit costs varying from \$105 to \$770 per tonne (\$95 to \$700 per ton). The values presented in table 5.1 are estimates and can vary substantially depending on the contaminant type and concentration, the media, and the quantity of material to be treated. The quoted unit costs include the cost of disposal and destruction or treatment of all residue, analyses associ-

**Table 5.1**
Cost Comparison

| Process | Quoted Costs $/tonne ($/ton)* | Wet vs. Dry Pricing Basis | Site Preparation Included | Quantity Tonne (Ton) | Disposal/ Destruction of Residues | Analytical | Mob**/ Demob | Profit Included |
|---|---|---|---|---|---|---|---|---|
| B.E.S.T. | 165 (150) | Wet | No | 18,000 (20,000) | Yes | Varies | Yes | Yes |
| CF Systems | 110-550 (100-500) | Wet | Yes | >57,000 (>63,000) | Yes | Yes | Yes | Unknown |
| Carver-Greenfield | 129-576 (117-523) | Wet | Yes | 21,000 (23,000) | Yes | No | Yes | Yes |
| Extraksol | 771 (700) | Wet | No | 910 (1,000) | Yes | Varies | Yes | Yes |
| LEEP® | 105-330 (95-300) | Wet | Yes | >40,000 (44,000) | Yes | Yes | Yes | Yes |
| NKD[a] | 138-330 (125-300) | Wet | No | 18,000 (20,000) | Yes | Varies | Yes | Unknown |
| S.R.U. | 220-661 (200-600) | Wet | No | 450 (500) | Yes | Varies | Yes | Yes |

* Costs are estimates only and are expected to be site specific.
** Mob = mobilization, demob = demobilization
a – data is for both process variations

ated with system operations (except for Carver-Greenfield), and mobilization and demobilization.

The US EPA has published detailed cost estimates for the CF Systems Process (US EPA 1990a) and the Carver-Greenfield Process (US EPA 1992f). These estimates include technology-specific costs and a breakdown of site-specific costs. In estimating costs for the CF Systems Process, the US EPA postulated the following scenarios:

- a base case treating 800,000 tonne (880,000 ton) of sediments contaminated with PCBs in concentrations of 580 ppm at 450 tonne/day (500 ton/day) over a 3.3-year period; and
- a hot-spot case treating 57,000 tonne (63,000 ton) of sediments contaminated with PCBs in concentrations of 10,000 ppm at 90 tonne/day (100 ton/day) over a one-year period; and
- analytical costs of $500/day in both of the above cases.

The estimated cost for the base case was $163±20% per tonne ($148 ±20% per ton) of raw feed, including excavation and pre- and posttreatment costs, but excluding final contaminant destruction costs. Excavation and pre- and posttreatment costs were estimated to be 41% of the total costs.

The estimated cost for the hot-spot case was \$492, -30% +50%, per tonne (\$447, -30% +50%, per ton) of raw feed. Excavation and pre- and posttreatment costs were estimated to be 32% of the total costs (US EPA 1990b).

The EPA's estimate for the Carver-Greenfield Process assumed treatment of 21,000 tonne (23,000 ton) of drilling mud contaminated with petroleum wastes. The total cost estimate was \$576/wet tonne (\$523/wet ton), with \$243/tonne (\$221/ton) allocated to technology costs. Site costs were estimated to be \$333/tonne (\$302/ton), including \$264/tonne (\$240/ton) for incineration of contaminated residuals. The estimate excluded regulatory, permitting, and analytical costs because of their variability. Excluded also were effluent treatment and disposal costs. Rather than assuming a cost for incineration, the vendor assumed that the process would separate indigenous oil, which would be sold to a refinery for \$26/tonne (\$24/ton), resulting in an overall cost of \$285/tonne (\$259/ton) (US EPA 1992g).

An estimate of the NKD Process treatment costs for the cleanup of a 15,300 $m^3$ (20,000 $yd^3$) site is in the \$164 to \$327/$m^3$ (\$125 - \$250/$yd^3$) range (Massey and Darian 1989). These costs were based on battery limits operation of a mobile system with net daily throughput rates in the range of 96 to 191 $m^3$/day (125 to 250 $yd^3$/day). An on-stream factor of 85% was assumed. The estimate included allowances for capital and operating expenses, waste disposal, mobilization, and demobilization.

For the treatment of refinery sludges, NKD originally set a treatment cost target of \$330/tonne (\$300/ton) of raw sludge. With a pilot study nearly complete, Massey and Darian (1990) reported that this target was still considered to be achievable. The estimate included allocations for pretreatment, amortization of capital, maintenance, manpower, chemicals, utilities, stabilization of solids, disposal of stabilized solids, and credit for recovered oil.

As the above discussion shows, costs for SCE can vary significantly depending upon the particular site. Treatability and site-specific cost studies are required to prepare meaningful cost estimates. The costs reported, however variable, appear competitive with costs for alternative remedial technologies.

# 6
# LIMITATIONS

## 6.1 Site/Matrix Considerations

Solvent/chemical extraction (SCE) processes are designed to remove and concentrate organic contaminants from soils, sludges, sediments, and wastewaters. Currently, they are not normally effective in removing inorganic contaminants, such as heavy metals. Although SCE processes have been demonstrated to be highly effective in the removal of polychlorinated biphenyls (PCBs) and polyaromatic hydrocarbons (PAHs), they may be less effective in treating organic compounds that are highly hydrophilic and of high molecular weight. The processes are not designed to treat particular compounds, and extraction efficiencies and processing rates are lower when there are high concentrations of indigenous organic compounds in the feed material. Similarly, extraction efficiencies and processing rates are lower when emulsifiers and water-soluble detergents are in the feed (US EPA 1991c).

Water was found to be a major impediment to effective operation of the NuKEM Development (NKD) soils remediation process. Early investigations revealed that as little as 10% moisture in the soil could significantly hinder extraction of PCBs. To overcome this problem, NKD has added a proprietary reagent (Massey and Darian 1989).

All of the SCE processes evaluated require some level of feed preparation before the extraction operation. Crushing and/or screening is required to reduce feed material to a maximum size of between 6 and 76 mm (0.25 and 3 in.). Most processes are limited to a well-delineated range of feed solids contents. This may necessitate a dewatering and drying stage in some processes and a slurrying stage in others. The Basic Extractive

Sludge Treatment (B.E.S.T.) Process may require pH adjustment to between 10.5 and 11, and the CF Systems Process, to between 6 and 10.

All of the processes addressed in this monograph have undergone substantial process development, testing, and evaluation. These evaluations have produced mixed results, often due to site-specific matrix considerations. In a number of cases, test objectives have been met or exceeded, while in other cases they have not been achieved. This reinforces the need for site-specific treatability testing and/or a detailed review of past test results.

## 6.2 Residue Treatment

The cleaned solids fraction may, in some cases, be deemed a Resource Conservation and Recovery Act (RCRA) hazardous waste, requiring stabilization or other treatment before disposal (see Section 3.7).

The SCE process can remove all indigenous organics and kill beneficial microbes in the feed material. If the residual solids are to be used as top soil, soil amendments will be required.

The concentrated contaminants fraction may often require additional treatment to ensure destruction of toxic organic compounds or to prepare this fraction for recycling (see Section 3.6). If organometallic compounds are present in the feed, they may be coextracted and concentrated in this fraction, thus further complicating final treatment. In some applications, however, this fraction may contain only petroleum hydrocarbons and, therefore, may be suitable for recycling as is.

The wastewater stream must be treated before being discharged to a water course and may require treatment before being discharged to a publicly owned treatment works (POTW) (see Section 3.7).

The extraction fluids can become contaminated with refractory compounds and require either a bleed stream or periodic replacement within the process. This results in another waste stream, which may require additional treatment before recycling or disposal.

## 6.3 Process Risks

All of the SCE processes use flammable organic extraction fluids presenting potential fire and explosion hazards. The flammability of these extraction fluids varies greatly. The CF Systems Process uses light pressurized hydrocarbons, presenting the greatest potential risk of explosion. Several of the extraction fluids include volatile or semivolatile compounds, which can create explosive vapor mixtures.

A number of the extraction fluids include toxic organic compounds, and therefore are subject to regulations governing the storage, use, transportation, and disposal of hazardous wastes. Process designs must minimize or eliminate personnel exposure to these compounds.

## 6.4 Reliability

SCE has only recently been applied in the remediation of contaminated soils, therefore, little data on commercial plant operations are available to evaluate long term reliability. Most data come from bench-scale, pilot-scale, or demonstration plants. U.S. Environmental Protection Agency Superfund Innovative Technology Evaluation (SITE) demonstration reports (US EPA 1990e; 1992a) for CF Systems and Carver-Greenfield identified some operating problems, including foaming of the extraction fluids, gumming-up of process lines, and intermittent sticking of solids to process equipment. Corrective actions have been identified, which, it is believed, will solve these problems in full-scale applications.

Although treatability tests at the bench scale have shown that SCE is applicable to a wide range of contaminants, they have shown also that process parameters must be optimized for each application.

More commercial applications treating run-of-site feed are needed to demonstrate whether SCE processes can handle the variations in feed properties that can be expected. Until these data are available, extensive site-specific treatability testing should be considered when applying this technology.

# 7 TECHNOLOGY PROGNOSIS

The use of solvent/chemical extraction (SCE) for the treatment of contaminated soils, sludges, sediments, and wastewaters represents new applications of a widely-used and well-understood technology. SCE is used in varied industries such as foods, pharmaceuticals, fire chemicals, mining, and minerals processing. The unit operations involved are simple and well understood.

Although the unit operations are well proven in other applications, their use for soil cleanup is still in its infancy. Most of the processes discussed in this monograph have few full-scale commercial applications. While it is expected that this will change in the future, SCE is a developing treatment technology requiring site-specific application testing and evaluation.

SCE has demonstrated a number of advantages in other industries. It is expected that these advantages will also apply to its use in treating soils, sludges, sediments, and wastewaters. These advantages include:

- Demonstrated high removal efficiencies and low residual values for a wide range of organic contaminants (polychlorinated biphenyls (PCBs), polyaromatic hydrocarbons (PAHs), petroleum hydrocarbons, pesticides, and dioxins);
- Demonstrated high concentration factors (up to 10,000:1), resulting in greatly reduced volumes of material requiring additional treatment; and
- The concentrated contaminant streams, especially when petroleum hydrocarbons are the soil contaminant, have the potential for recycle.

Although SCE has had limited application to date, there are indications pointing to its likely expanded application in site remediation, including the following:

- Several SCE processes are under, or have completed, evaluation in the U.S. Environmental Protection Agency's (US EPA's) Superfund Innovative Technology Evaluation (SITE) demonstration program, which provides independent verification of the processes' efficiency, operability, and cost;
- Commercial SCE processes are already being used to treat petroleum refinery and other waste streams, allowing determination of long-term costs and system reliability;
- The SCE processes do not require extensive pretreatment of the feed (other than size reduction) and can tolerate a wide range of soil moisture content (from about 5% to 90% moisture); and
- The SCE processes are cost-competitive with other technologies used to treat organic-contaminated soils, sludges, and sediments.

Several companies are working on developing proprietary SCE processes that will expand options for applying SCE (see Appendix B for discussions of some emerging processes). Further, as additional systems are brought on line, process uncertainties may be minimized and treatment costs should decrease.

# A

# VENDOR CONTACTS

### B.E.S.T.

Resource Conservation Corporation
3006 Northrup Way
Bellevue, Washington 98004-1407
Phone: (206) 828-2400

### LEEP

ART International, Inc.
100 Ford Road
Denville, New Jersey 07831
Phone: (201) 627-7601

### CF Systems

Morris Knudsen Company
3D Gill Street
Woburn, Massachusetts 01801
Phone: (617) 937-0800

### NKD Process

NuKEM Development
3000 Richmond Avenue
Houston, Texas 77479
Phone: (713) 520-9994

### CF Technologies

1 Westinghouse Plaza, Suite 200
Hyde Park, Massachusetts 02136-2059
Phone: (617) 364-2500

### SRU

Terra-Kleen Corporation
7321 North Hammond Avenue
Oklahoma City, Oklahoma 73132
Phone: (405) 728-0001

### Carver-Greenfield Technology

Dehydro-Tech Corporation
6 Great Meadow Lane
East Hanover, New Jersey 07936
Phone: (201) 887-2548

### Contex Process

A/S Phoenix Miljø
Vester Alle 1
6600 Vejen, Denmark
Phone: (45-75) 361-111

### Extraksol Process

CET-Sanivan
7777 Boulevard L.H. Lafontaine
Anjou, Quebec H1K 4E4, Canada
Phone: (514) 355-3351

### LIX Process

Henkel Corporation
2330 Circadian Way
Santa Rosa, California 95407
Phone: (707) 576-6227

### RIP/CIP Process

Davy Research & Development
P. O. Box 37
Bowesfield Lane
Stockton-on-Tees, Cleveland TS18 3HA
England
Phone: (44-642) 607-108

# B
# EMERGING PROCESSES

In addition to the processes addressed in the monograph proper, several other processes that show promise for expanding the use of solvent/chemical extraction (SCE) in site remediation are being developed. Five processes, demonstrated at the bench scale, were deemed to merit review here (see table B.1 on page B.2). Two of them, the Liquid Ion Exchange (LIX) and Resin-in-Pulp/Carbon-in-Pulp (RIP/CIP) Processes, incorporate traditional mining and minerals processing concepts in separating inorganic, as well as organic, contaminants from soils. One of the processes, Contex, has been employed commercially in Denmark and is currently undergoing major modifications to improve throughput. The Adiabatic Process for the Extraction of Sludges and Soilex Process have been successfully tested at the bench scale, but further development has not been undertaken.

*Chemical Waste Management Adiabatic Process for the Extraction of Sludges.* In the late 1980's, Chemical Waste Management began developing the Adiabatic Process for the Extraction of Sludges, at times called the APES, for removing water and organics from sludges (Henry and Gillenwater 1992). Bench-scale studies were carried out, and construction of a pilot plant was begun. These efforts were discontinued in 1991.

Figure B.1 (on page B.3) is a flow schematic for a scenario in which a sludge containing solids, water, and relatively volatile organics is to be treated with the objective of producing dry, organic-free solids and recovering at least part of the organic contaminants. The process is to be fully continuous. In the evaporation step, sludge is fed into a forced circulation evaporator in which a heavy, "nonvolatile" oil is circulated. When the heavy oil and sludge contact moisture, volatile organic compounds and semivolatile organic compounds are evaporated. These vapors are sent forward to a catalytic oxidation system in which the organic compounds are destroyed. The catalytic oxidation step is not required for sludges that do not contain volatile organics.

**Table B.1**

Bench Scale & Promising SCE Processes

| Technology Name | Developer | Level of Development | Targeted Applications |
|---|---|---|---|
| APES | Chemical Waste Management | Bench scale.<br>Pilot plant constructed. | Organic contaminant removal from sewage, millscale, and petroleum sludges. Not currently marketed. |
| Contex | Phoenix Miljø | Full-scale fixed facility, 9 tonne/hr (10 ton/hr).<br>Process undergoing major modification.<br>Restart Dec. 1992.<br>Methylene chloride solvent. | Organic contaminated soils. +99% removal of BTX, naphthalene, phenanthrene, and chlorinated solvents reported. |
| LIX | Henkel | Bench scale.<br>Utilizing traditional mining SX processes. | Heavy metals (Cd, Cu, Ni, Pb, Zn) removal from soils. |
| RIP/CIP | Davy Research & Development | Bench scale.<br>EPA SITE program.<br>Traditional mining processes. | Inorganic contaminant removal using resin-in-pulp (RIP) technology. Organic contaminant removal using carbon-in-pulp (CIP) technology. |
| Soilex | Martin Marietta | Bench scale and batch pilot tests, 1984, DOE Oak Ridge, Tennessee. | PCB removal from soils. +90% removal reported. Not currently marketed. |

A second stream leaving the evaporator is composed of solids, the heavy oil utilized in the evaporator, and any low-volatility oils in the original sludge feed. In the extraction step, a volatile alkane solvent is used to separate the oils from the solids in a continuous countercurrent multistage operation.

The extract, which contains the light solvent and the oils, is then fractionated to recover the solvent; the solvent is recycled. The bulk of the oil is recycled to the evaporator. Any excess beyond that needed to maintain the inventory in the evaporator is disposed of.

The solvent-laden solids from the extraction step are sent to a dryer. This is the *solids desolventization phase*, part of the product preparation step. Solvent volatilized in the dryer is recycled, and the dry, oil-free solids are discharged.

**Figure B.1**
Chemical Waste Management APES Flow Schematic

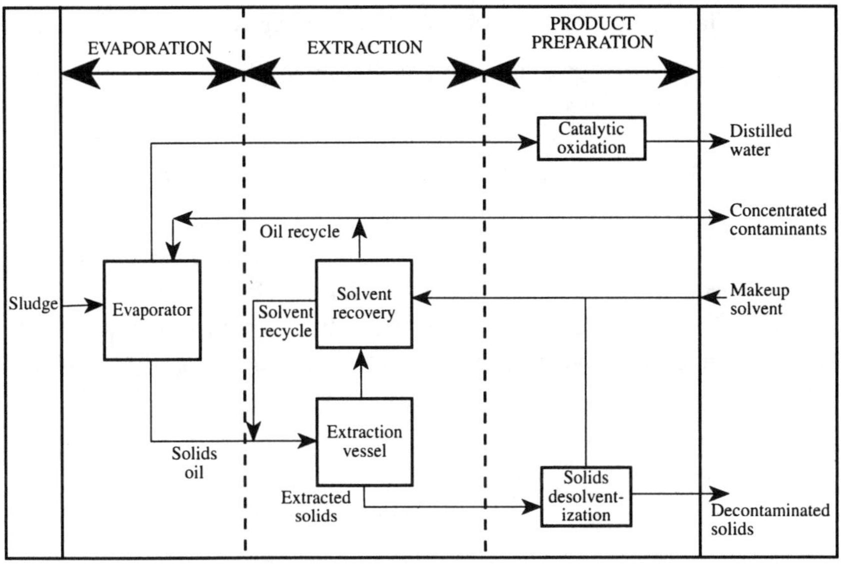


Intended applications included treating municipal sewage sludge, mill-scale sludge, and petroleum refinery sludges (Resource Conservation and Recovery Act (RCRA) waste codes K048 through K052).

*Phoenix Miljø Contex Process.* Phoenix Miljø (Environmental) of Vejen, Denmark (Phoenix), has developed and operated a semimobile SCE process for the treatment of contaminated soils from industrial sites and drilling operations (Nørregaard 1992). In 1990, Phoenix built and began operating a nominal 9-tonne/hr (10 ton/hr) demonstration plant at its facilities and has tested between 20 to 30 different soils contaminated with tar, petroleum hydrocarbons, drilling cuttings, chlorinated solvents, benzene, toluene, xylene (BTX) compounds, naphthalene, and phenanthrene. The vendor claims very high removal efficiencies and low residual soil organics in their brochure (Phoenix 1991).

Liquid-solid extraction is combined with steam stripping in a process whose operating parameters can be adjusted to achieve required removal efficiency for each application (see figure B.2 on page B.4). Contaminated soil is fed through an inlet seal leg into a series of screw conveyors. The extraction fluid, methylene chloride, is fed countercurrent to the soil.

The first conveyor serves as the extraction stage. The second conveyor serves as a drainage device to remove extraction fluid from the soil. The soil then enters a steam-stripping unit, where residual extraction liquid is removed. The decontaminated soil leaves the extraction system via an outlet lock.

The extraction liquid containing the organic contaminants is condensed and transferred to the extraction conveyor. Makeup lean extraction fluid from a buffer tank is added at this point. The loaded extraction liquid flows from the extraction conveyor to a distillation unit, where it is separated for

**Figure B.2**

Phoenix Contex Process

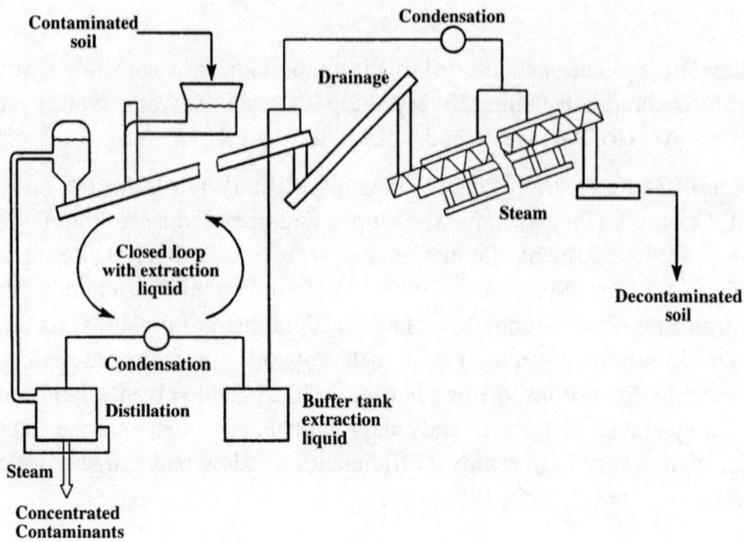

recycle. The concentrated contaminant stream is collected for transport to an off-site facility for further treatment.

Phoenix is currently performing major modifications to its on-site processing plant. The goals of these modifications are to improve system reliability and increase plant capacity for a wider range of soils. Previous experience has been that clay-containing soils cause handling problems and reduce plant capacity. The modifications were scheduled to be completed in mid-1993, at which time characterization testing will resume.

The developer indicates very favorable economics for the Contex process. A very rough estimate of the capital costs for the demonstration plant is between $2,000,000 and $2,300,000. Treatment costs of $110 to $193/tonne ($100 to $175/ton) were quoted. Significant feed properties that influence treatment costs are soil water and oil content, as well as soil type (Nørregaard 1992).

*Henkel Liquid Ion Exchange Process.* Henkel Corporation is a major producer of liquid ion exchange reagents for the mining and minerals processing industry. These reagents are widely used to extract metals, such as copper, nickel, uranium, and zinc, from leaching solutions having great specificity for metals of interest in liquids that contain high concentrations of several competing compounds.

Henkel's Hazardous Waste Division has performed a series of bench-scale treatability studies of soils contaminated with heavy metals (Virnig 1992). To date, the work has concentrated on the removal of lead from soils. Additional work is planned on developing process concepts and testing Liquid Ion Exchange reagents for the removal of cadmium, copper, nickel, and zinc from contaminated soils.

Successful demonstration of the Henkel LIX processes may lead to a much wider application of SCE in the treatment of contaminated soils.

*Davy R&D Resin-in-Pulp/Carbon-in-Pulp Process.* Davy International's Research and Development Division (Davy R&D) in Stockton, England, is developing an RIP/CIP Process for the treatment of contaminated soils. The RIP/CIP technology is used in the mining and minerals processing industry to concentrate valuable metals from slurried ore bodies and their leachates. The RIP/CIP Process is suitable for the treatment of a wide range of materials contaminated with both inorganic and organic wastes (US EPA 1992e).

**Figure B.3**
Davy R&D RIP/CIP Process

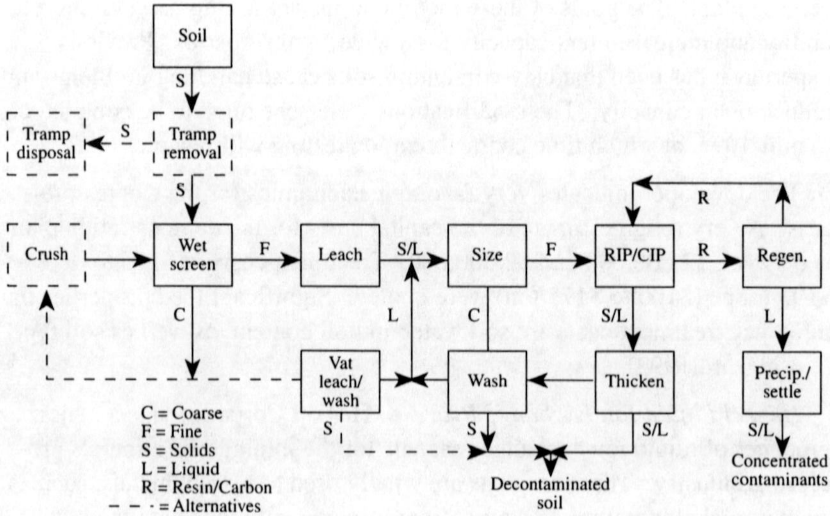


The process (see figure B.3 on page B.6) entails leaching of organic and heavy metals from contaminated soils followed by the adsorbing and removing of contaminants by activated carbon or ion exchange resins. The RIP ion exchange processes are most appropriate for metal removal, whereas the CIP adsorption processes are more suitable for the removal of organic contaminants.

Contaminated material is passed through screening stages in order to separate tramp and oversize material. In the case of metal contamination, the oversize material could be crushed or undergo a vat leach before washing and disposal. The soil then passes to stirred vessel leaching stages. The leached pulp is physically separated into a sand fraction and a slime fraction. The sand fraction wash liquor (the leachate), and the slime fraction (the pulp) contain the bulk of the contaminants. The pulp is passed to a multistage RIP contactor, where the contaminants are removed by ion exchange resin. The resin is stripped to remove the contaminants and then recycled back to the RIP contactor. The concentrated contaminants present in the strip solution are precipitated as a concentrated sludge for recycling or disposal.

A similar process will be used in removing organic contaminants, with appropriate aqueous reagents and surfactants used as leaching agents. The activated carbon will cause the contaminants to be desorbed chemically or thermally, and the carbon will be reactivated for reuse in the CIP process. Desorption products can be disposed of or recycled.

The technology was accepted in the Superfund Innovative Technology Evaluation (SITE) Emerging Technology Program in July 1991, and laboratory bench-scale tests have been underway since (US EPA 1992e). Initial efforts focused on remediation of soil contaminated with arsenic, copper, and chromium. These tests have shown removal efficiencies of over 90%. More recent tests have concentrated on removing mercury from contaminated soils.

Process data from the above tests are being used to prepare conceptual cost estimates (± 30%) for a nominal 140-tonne/day (150 ton/day) plant. Locations were being screened for operation of a 1.8-tonne/day (2 ton/day) pilot plant beginning sometime in 1993. Process data are being developed for treating soils containing both metals and organic contaminants (Naden 1992).

*Martin Marietta Soilex Process.* Martin Marietta Energy Systems, Inc., (MMES) conducted bench- and pilot-scale studies of extraction of polychlorinated biphenyls (PCBs) from soil in 1984 using their Soilex Process (Napier, Hancher, and Saunders 1990; Hancher, Napier, and Kosinski 1984), but no attempt to commercialize the technology has been made or is planned. The studies were performed at the U.S. Department of Energy Y-12 Plant in Oak Ridge, Tennessee, on soil containing PCBs in concentrations of 300 to 600 ppm.

In the process shown in figure B.4 (on page B.8), feed was reduced in size and screened so that most soil particles were less than 6 mm (.25 in.) in diameter. The screened soil was then washed with a mixture of 50% water and 50% kerosene by volume. The PCBs and oil were extracted into the water/kerosene phase, the soil was dried and chemically analyzed, and the extent of PCB removals was determined.

The water and kerosene mixture was allowed to separate in phase, and the water was returned to the process. The kerosene extract was distilled to obtain purified kerosene, which was returned to the process, and a PCB/oil-contaminated phase. The kerosene used in these studies had a boiling point of 180° to 210°C (356° to 410°F), a flash point of 50°C (122°F), and an auto-ignition temperature of 360°C (680°F).

Before the pilot-scale studies, a number of laboratory shakedown tests were conducted to define the best water-kerosene-soil ratios. Satisfactory

**Figure B.4**
Martin Marietta Soilex Flow Schematic

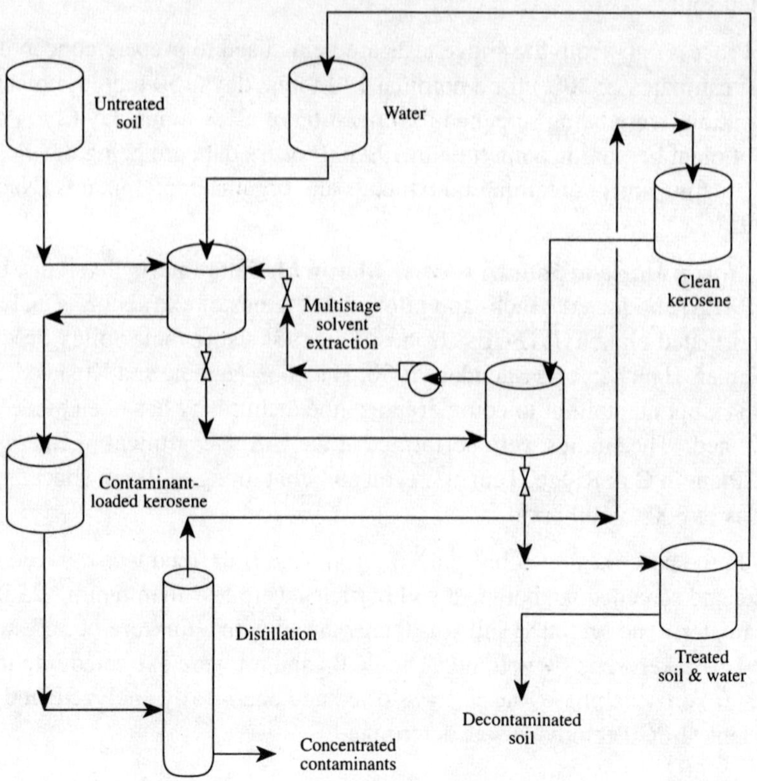

results were obtained using equal volumes of water and kerosene plus 10% soil by weight. In one test, soil containing a concentration of 600 ppm PCBs underwent three extractions and the concentration was reduced to 13 ppm, an overall reduction of 97.8%. The rate of removal was nominally 70% per stage.

The three-stage countercurrent extraction pilot-plant operation was a batch process with each stage requiring a nominal 24 hours for mixing and settling. Each extraction stage employed a 200-L (55 gal) drum equipped with an agitator to stir the soil, water, and kerosene mixture. The soil and water were added in the first stage, and clean kerosene was added in the third stage. Either 15 or 24 kg (33 or 53 lb) of contaminated soil was added in the first stage, which contained 76 L (20 gal) of water and 76 L (20 gal) of kerosene from the second stage. The mixture was stirred for about 30 minutes and then allowed to settle for 16 hours. The treated soil was then sent to the second stage, etc. The three extraction stages required three days to complete, but once the stages were filled, a batch was added and taken from the process daily.

Seventeen runs were made in the pilot plant, and the three extraction stages removed an average of 91.6% of PCBs over a removal range of 85.3% to 97.9%. The results indicate that each additional stage would have removed about 70% of the remaining PCBs. Two tests reduced the PCBs from starting concentrations of 280 and 340 ppm to 6 ppm. After three stages, the PCB range of the treated soil from all of the pilot-plant tests was 6 to 50 ppm, in soil contaminated initially with 240 to 360 ppm PCBs.

The kerosene phase (extract) from the first stage was distilled to obtain a concentrated PCB-kerosene phase and a recyclable kerosene containing practically no PCBs. In laboratory tests, 87.5% of the kerosene could be vacuum distilled and used in the recycle operation. The pilot-plant distillation was carried out in a steam-heated still that could not be operated in a vacuum mode. Because of the low-pressure steam used in the plant, a maximum temperature of 150°C (302°F) was applied, allowing recovery of only 82% of the kerosene. Higher distillation temperatures would have allowed more kerosene to be recycled, reducing the amount of PCB-kerosene left for disposal by incineration.

The water/kerosene phase separated during the soil settling period. The kerosene was transferred to the next extraction stage or to the distillation process. The water phase was separated from the soil after the third extraction stage through a settling stage, and the water was returned to the first stage without purification. The water was used to reduce the amount of kerosene required and to serve as a fluid to transfer the soil from one wash station to the next.

The soil was wet with water and kerosene after the third stage. The pilot-plant tests did not remove either the water or the kerosene, but they would have been removed through heating the soil and condensing the liquids if the process had been further developed.

Pilot-scale tests were conducted of a process to extract PCBs from oil employing a continuous countercurrent extraction column and using dimethyl formamide as the solvent. After extraction, water was added to the dimethyl formamide and PCB extract, causing a phase separation with one phase containing the PCBs and the other, the dimethyl formamide and water. The latter mixture was then separated by distillation, and both products were recycled through the process. As with the PCB-in-soil process, no attempt has been made to commercialize this technology. (Hancher, Napier, and Kosinski 1984.)

# C

# CARVER-GREENFIELD PROCESS

Results of four studies of the Carver-Greenfield Process treating a variety of contaminated materials are reported below. The pilot-scale data in case study #1 were provided in the report of an U.S. Environmental Protection Agency (US EPA) Superfund Innovative Technology Evaluation (SITE) demonstration. The remaining studies were conducted by Dehydro-Tech Corporation (DTC) and were not available for review.

## CASE STUDY #1: LABORATORY TESTS AND US EPA SITE PILOT-SCALE DEMONSTRATION AT PAB OIL

The US EPA SITE demonstration was conducted in August, 1991 using drilling mud waste from the PAB Oil and Chemical Services (PAB Oil) Superfund Site in Abbeville, Louisiana (US EPA 1992a). Drilling mud wastes consist of oil, solids, and water that become mixed during the production of oil (US EPA 1992a). As reported by the developer, initial laboratory-scale tests showed the wastes consisted of 29.5% solids, 29.1% water, and 41.4% oil (Trowbridge, Holcombe, and Kollitides 1991). The laboratory tests reduced toluene content from 18,600 ppb to less than 350 ppb, while acetone, ethylbenzene, and xylenes, which had been present in the feed in concentrations of from 2,000 to 10,000 ppb, were undetectable in the treated solids (Environment Today 1992).

Following the laboratory tests, a 44-kg/hr (100 lb/hr) pilot-scale system was used in the 1991 SITE demonstration to assess how effectively the process separated petroleum-based, hydrocarbon-contaminated drilling mud wastes into their solid, oil, and water fractions. The study included shakedown runs (to optimize operating conditions), a blank run (with no waste treatment), and two test runs. The following information is based primarily on the US EPA Aug. 1992 Applications Analysis Report (US EPA 1992a).

Isopar-L was used as the extraction solvent, and the waste was screened to 0.64 cm (0.25 inch) before processing. The raw feed and solids product were analyzed for solids/indigenous oil/water (SOW) content. Although the raw feed contained significant levels of indigenous oil and elevated levels of metals (US EPA 1992a), the feed, as well as the treated solids, passed toxicity characteristic leaching procedure (TCLP) tests.

During shakedown testing treating silt, a gummy material formed, plugging process lines. In the next shakedown, a surfactant was used with the silt to prevent formation of the gummy substance. The final product, however, was found to contain 4% indigenous oil, while the clean silt contained no indigenous oil. The report concluded that the surfactant had been detected as indigenous oil in the SOW analysis, and its use was, therefore, discontinued (US EPA 1992a).

A blank run was then conducted using bentonite as the solid matrix, and no surfactant was added. Once again, lines became blocked with gummy material, and this initial blank run was discontinued. It should be noted that different materials were used in the shakedown and blank runs; therefore, consistent process adjustments could not be made.

Following these runs, the SITE demonstration was started. A blank run using bentonite was conducted, and waste feed processing began. Two waste feed test runs, each consisting of three extractions, were conducted. The system was operated in a batch mode using fresh solvent for each extraction, because of the pilot-scale nature of the test (US EPA 1992a). Samples of waste feed, raw solvent, slurried feedstock, concentrate (from each extraction), centrifuge cake, condensed water, condensed solvent, solids product, and vent gas were collected. The condensed water and condensed solvent resulted from materials collected from evaporation during the extraction steps that were condensed and then manually separated using a separatory funnel. The vent gases were those that had not been condensed, and were passed through a granular-activated carbon canister (US EPA 1992d).

Approximately 136 kg (300 lb) of feed was fluidized in 680 kg (1,500 lb) of solvent (Trowbridge 1992). As the developer explained, the extraction temperature of 66° to 93°C (150° to 200°F), was maintained for 15 minutes to one hour. In one of the extractions, the mixture was heated to a higher temperature (around 107° to 135°C (225° to 275°F)) to evaporate the water from the feed. After centrifuging, the solvent was vaporized from the solids using inert gas at 177°C (350°F) (Trowbridge, Holcombe, and Kollitides 1991).

The analytical results of the tests are presented in table C.1 (on page C.4). Each feedstock contained approximately 52% solids, but the indigenous oil content differed (feedstock 1 was ≈17% oil, and feedstock 2 was ≈7% oil). Based on summaries of the SITE demonstration (US EPA 1992f; US EPA 1992a), the final solids product contained approximately 0.9% carrier oil (0.93% for test 1, and 0.89% for test 2) and approximately 0.9 to 1.5% indigenous oil (1.38 to 1.45% for test 1 and 0.85% for test 2).

The method of determining oil removal made it difficult to characterize test results. The Isopar-L solvent is detected through total petroleum hydrocarbon (TPH) analysis. Consequently, the SITE demonstration data was based on concentrations calculated by subtracting Isopar-L levels (based on gas chromatography to determine the Isopar-L levels). The following tests were used to gather data on oil:

- Solids/Indigenous Oil/Water Content. The indigenous oil detected through this analysis includes a broad range of polar and nonpolar organics that are soluble in toluene, including petroleum hydrocarbons. Polar organics or surfactants may be detected as oil in the SOW procedure;
- Total Petroleum Hydrocarbon. The TPH analysis uses silica gel, which contains polar organics and surfactants (Trowbridge 1992). As a result, the TPH analyses detected only nonpolar organics, including the Isopar-L solvent; and
- Solvent Levels. The Isopar-L solvent levels were characterized through gas chromatography.

The SITE demonstration calculated "indigenous TPH" by subtracting the Isopar-L result from the TPH result. This calculation was then used to determine removal efficiencies. Several important qualifications are introduced by the SITE demonstration method of calculating removals. First, the resulting calculations sometimes resulted in negative concentrations (the Isopar-L gas chromatographic results were higher than the total TPH levels, resulting in a negative calculation for the indigenous TPH). The indigenous TPH was therefore reported as 0 in the final product, resulting in calculations of 100% removal (see table C.2 on page C.6). This approach may be flawed, however, since it is based on the assumption that the obvious discrepancies in the analyses (perhaps due to nonhomogenous sample results, analytical difficulties in quantifying exact levels, or other factors) act solely

## Table C.1
Carver - Greenfield Site Demonstration

| Parameter | Test Run #1 | | | | Test Run #2 | | | |
|---|---|---|---|---|---|---|---|---|
| | | Final Products | | | | Final Products | | |
| | Feedstock | Solids | Water | Solvent | Feedstock | Solids | Water | Solvents |
| SOW | (%) | | | | | | | |
| - Solids | 52.35 | 96.42 | NA | NA | 52.44 | 98.31 | NA | NA |
| - Indigenous Oil | 17.48 | 1.38 | NA | NA | 7.24 | 0.85 | NA | NA |
| - Water | 21.75 | <0.1 | NA | NA | 34.77 | <0.1 | NA | NA |
| VOC | μg/kg | | | | | | | |
| - Benzene | -- | -- | -- | -- | 1075.67 | NA | <250 | NA |
| - Toluene | 546 | NA | <250 | NA | 1046.00 | NA | <250 | NA |
| - Ethylbenzene | 993 | NA | <250 | NA | 1886.67 | NA | <250 | NA |
| - Total Xylene | 3,658 | NA | <250 | NA | 8,873.33 | NA | <250 | NA |
| - Acetone | ND | NA | 4,927 | NA | <5,000 | NA | 2,280 | NA |
| - 2-Butanone (MEK) | ND | NA | 1,067 | NA | <2,500 | NA | <395.67 | NA |
| SVOC-acid | μg/kg | | | | | | | |
| - Phenol | <100,000 | <660 | <203 | NA | ND | ND | ND | NA |
| SVOC - Base Neutral | lg/kg | | | | | | | |
| - Phenanthrene | 15,950 | 257 | <250 | NA | 8,126.67 | <1,650 | <50 | NA |
| - 2-Methylnaphthalene | <26,183 | <660 | <500 | NA | 49,150.00 | 2,316.67 | <100 | NA |
| - Isophorone | <50,000 | <364 | <250 | NA | -- | -- | -- | -- |
| - bis(2-ethylhexyl)-phthalate | <50,000 | 592 | <250 | NA | <50,000 | 1,406.67 | <196.27 | NA |
| - Di-n-oxtyl pthalate | <50,000 | <321 | <250 | NA | <50,000 | ND | <20.1 | NA |
| - Napthalene | -- | -- | -- | -- | <28,417 | 1,054.67 | <50 | NA |

| Metals | µg/g | | | | | | | |
|---|---|---|---|---|---|---|---|---|
| - Aluminum | 10,663 | 16,833 | 261 | 11.43 | 7,351.67 | 8,606.67 | 13.33 | <9.94 |
| - Antimony | <5.0 | <5.0 | <0.052 | <5.0 | -- | -- | -- | -- |
| - Barium | 2,990 | 3,193 | 0.0382 | <0.50 | 575.83 | 3,173.33 | <0.50 | <0.50 |
| - Beryllium | 0.831 | 1.37 | <0.0050 | <0.50 | 0.70 | 1.12 | <0.50 | <0.50 |
| - Boron | <24.7 | <20 | <0.20 | <20 | <21.93 | ND | <20.0 | <24.33 |
| - Cadmium | 0.578 | <0.50 | <0.0050 | <0.50 | 4.00 | 5.55 | <0.50 | <0.50 |
| - Calcium | 2,135 | 5,243 | 4.28 | <50 | 7,785.00 | 12,033.00 | <52.37 | <48.27 |
| - Chromium | 25.4 | 412 | <0.0103 | <1.0 | 139.50 | 318.33 | <1.0 | <1.0 |
| - Cobalt | 7.43 | 14.8 | <0.050 | <5.0 | 9.41 | 13.23 | <5.0 | <5.0 |
| - Copper | 16.4 | 26.4 | <0.010 | <1.0 | 88.50 | 107.33 | <1.0 | <1.0 |
| - Iron | 13,567 | 20,333 | 2.68 | <5.0 | 20,733.00 | 36,433.00 | <6.46 | <5.0 |
| - Lead | 41.0 | 46.1 | <0.050 | <5.0 | 205.17 | 248.33 | <5.0 | <5.0 |
| - Magnesium | 1,517 | 2,460 | 1,085 | <50 | 1,251.67 | 1,926.67 | <50 | <50 |
| - Manganese | 373 | 695 | 0.0933 | <0.5 | 276.00 | 505.33 | <0.5 | <0.5 |
| - Molyodenum | <5.0 | 43.3 | <0.050 | <5.0 | 25.35 | 48.97 | <5.0 | <5.0 |
| - Nickel | 13.3 | 43.3 | <0.0384 | <2.0 | 20.83 | 145.67 | <2.0 | <2.0 |
| - Potassium | 485 | 287 | <0.56 | <50 | 747.17 | 969.33 | <50 | <50 |
| - Sodium | 135 | 1310 | 3.05 | <50 | 599.17 | 2,310.00 | 67.97 | <53.73 |
| - Strontium | 64.5 | 125 | 0.0576 | <5.0 | 270.67 | 378.00 | <5.0 | <5.0 |
| - Vanadium | 24.3 | 28.0 | <0.050 | <5.0 | 22.03 | 21.27 | <5.0 | <5.0 |
| - Zinc | 160 | 145 | 0.192 | <1.0 | 1,013.33 | 1,153.33 | <1.0 | <1.21 |
| SOLVENT | (%) | | | | | | | |
| - Isopar I | <0.1 | 0.88 | 0.900 | 96.07 | <0.1 | 0.99 | 0.10 | 99.70 |
| TPH | (µg/g) | | | | | | | |
| - TPH | 146,833 | 7,907 | 1,442 | 952,667 | 89,383 | 6,617 | 333 | 923,333 |

SOW = solids/indigenous oil/water
VOC = volatile organic compound
SVOC = semivolatile organic compound
ND = not detected
NA = not analyzed

Source: US EPA 1992a

**Table C.2**
Carver-Greenfield Site Demonstration: Oil Parameters

| Parameters | Test Run #1 | | Test Run #2 | |
|---|---|---|---|---|
| | Feedstock | Final Solids | Feedstock | Final Solids |
| SOW[1] | | (%) | | |
| - Solids 5 | 2.35 | 96.56 | 52.44 | 98.31 |
| - Indigenous Oil | 17.48 | 1.38 | 7.24 | 0.85 |
| - Water | 21.75 | 0 | 34.7 | 0 |
| TPH[1] | (%) | | | |
| -TPH | 14.7 | 0.79 | 8.9 | 0.66 |
| Isopar - L[1] | (%) | | | |
| - as solvent 0 | 0.93 | 0 | 0.99 | |
| - as TPH 0 | 0.84 | 0 | 0.89 | |
| Calculated Oil[1] | (%) | | | |
| - indigenous oil | 17.47 | 1.38 | 7.24 | 0.85 |
| - indigenous TPH | 14.7 | 0 | 8.9 | 0 |
| - non-TPH oil | 2.77 | 1.38 | -- | 0.85 |
| Oil Removal Efficiency[1] | (% removal) | (% removal) | | |
| - true indigenous oil | 92.1 | 88.3 | | |
| - indigenous TPH | 100 | 100 | | |
| Oil Removal Efficiency[2] | (% removal) | (% removal) | | |
| - oil | 95.9 | 94.3 | | |

1. Source: US EPA 1992a, C-11
2. Source: US EPA 1992f. These % removals were reported in a SITE demonstration bulletin, based on the solids fraction of the influent feed. The bulletin, which summarized results of the SITE demonstration (US EPA 1992a), reported test run #1 indigenous oil of 1.45% (instead of 1.38%) and carrier oil of 0.93%. Test run #2 indigenous oil results (0.85%) matched reference 19, and carrier oil was noted to be 0.89%.

in favor of assuming 100 percent removal, without quantifying a potential range. Second, simply subtracting the Isopar-L solvent levels from the TPH analysis results ignores the fact that TPH analysis may itself be important (for regulatory purposes) in determining whether a product is clean. Therefore, the actual TPH rather than the calculated indigenous TPH may be a more important factor in evaluating whether the process can be considered to have successfully treated a material (the fact that the final solids contain Isopar-L, which is detected as TPH, must be considered).

As explained in the SITE demonstration report (US EPA 1992a), the SITE approach qualitatively (rather than quantitatively) characterizes oil removal, based on a treatment objective of removing indigenous oil.

A review of table C.1 (on page C.4) reveals that TPH removals (based on total TPH, not indigenous TPH) were 94.6% for test run 1 (146,833 reduced to 7,907 μg/g) and 92.6% for test run 2 (89,383 reduced to 6,617 μg/g). As described earlier, indigenous TPH removals were calculated to be 100%. Oil removals based on the solids fraction of the influent feed were ≈ 94 to 96% (US EPA 1992g), but indigenous oil removals were ≈ 88 to 92% (US EPA 1992d).

A gross mass balance (of solids, water, and oil) showed overall recoveries of about 96% for both runs. Specifically, test run 1 showed recoveries of 81.2% of solids, 112.2% of water, and 97.3% of oil; and test run 2, 78.9% of solids, 95.8% of water, and 98.3% of oil (US EPA 1992a).

The following conclusions were stated:

- The Carver-Greenfield Process removed approximately 90% of the indigenous oil and reduced indigenous TPH to trace levels on the solids product (US EPA 1992a);
- The solids product consists of a dry powder, similar to dry bentonite with residual oil, attributed primarily to the Isopar-L solvent (a food grade oil) (US EPA 1992a). It should be noted that other US EPA documents reported approximately equal final levels of indigenous oil and carrier oil (test run 2, 0.85% and 0.89%), or even higher indigenous oil levels than carrier oil levels (test run 1, 1.45% and 0.93%) (US EPA 1992g);
- Metals and organics met TCLP criteria, but wastes may, nonetheless, have to be disposed of as hazardous materials because of regulatory requirements;
- The process does not remove metals bound to the solids and may increase the apparent metals concentration in the solids because of volume reduction (US EPA 1992d). Because of this effect, as table C.1 (on page C.4 - C.6) shows, the metals concentrations in the final product solids were consistently higher (on a μg/g basis) than those in the influent feed, except for boron, potassium, and zinc (test run 1) and boron and vanadium (test run 2). Most of the reported metal concentrations, however, were below de-

tection limits. In addition, the US EPA noted that increases in TCLP extracts could be expected because of the proportional increase in the amount of solids in the final product, although there was no evidence that the process increased the actual leachability of metals (US EPA 1992a); and

- The effluent water requires further treatment because of the presence of light organics and solvent, but may in some instances be discharged to a local publicly owned treatment works (POTW) (US EPA 1992a).

The average TPH, based on two test runs, each of which was treated with three extractions, was reduced from 11.8% in the feed to 0.8% in the final solids (Holcombe, Trowbridge, and Rawlinson 1991). See table C.3.

Table C.4 (on page C.9) presents a summary of the cost analysis in this case study. The difference between the US EPA estimate ($244.00/tonne ($221.42/ton)) and the vendor estimate ($122.38 to 238.68/tonne ($111.05 to $216.59/ton)) for the technology-specific cost elements is due primarily to different assumptions about the amount of labor required. Data from commercial-scale operations should dispell the disparity.

**Table C.3**
Carver-Greenfield Site Demonstration: Average Compositions By Weight Percent

| Component | Feedstock | Treated Solids |
|---|---|---|
| Solids | 52.4 | 97.4 |
| Water | 28.2 | <0.1 |
| "Food-Grade" Solvent | 0.0 | 1.0 |
| Other (non-volatile toluene extractables) | 12.4 | 1.1 |
| Total | 93.0 | 99.6 |
| TPH | 11.8 | 0.8 |

Reprinted by permission of Theodore D. Trowbridge from "The Carver-Greenfield Process for the Treatment of Oily Refinery Sludges" by Thomas Holcombe, Theodore D. Trowbridge, and Charles T. Rawlinson, presented at the Third Annual CERT Symposium, "Environmental Protection in the Energy Industry: Treatment and Disposal of Petroleum Sludges", University of Tulsa, December 12, 1991. Copyright 1991 by Theodore D. Trowbridge.

**Table C.4**
Carver-Greenfield Cost Analysis Summary

| Cost Category | EPA Estimates[1] (Site Demo) | Vendor Estimates[1] (Site Demo) |
|---|---|---|
| Technology-Specific | | |
| Equipment | $21.85 | $16.16 to $30.18 |
| Start-up and Fixed | 4.47 | 5.48 |
| Labor | 163.27 | 59.05 to 147.62 |
| Supplies (Isopar-L solvent) | 8.90 | 8.90 |
| Consumables | 15.36 | 15.31 |
| Facility Maintenance | 4.79 | 3.41 to 6.36 |
| Demobilization | 2.78 | 2.74 |
| Technology-Specific Subtotal | $221.42 | $111.05 to $216.59 |
| Site-Specific | | |
| Site Preparation/Excavation | 55.00 | 30.22 to 53.96 |
| Residuals Treatment | 247.02 | -23.90 to +7.30 |
| Site-Specific Subtotal | $302.02 | $6.32 to $61.26 |
| TOTAL | $523.44 | $117.37 to $246.65 |

1. Per wet ton of drilling mud waste (31% water, 17% oil, 52% solids) treated, assuming treatment of 23,000 tons of waste

US EPA 1992a

The large difference between the US EPA's and vendor's estimates of site-specific costs is attributable to different assumptions as to the disposition of residues. The US EPA posits a cost for treating residue, whereas the vendor assumes significant recovery and reuse of the residues, thereby offsetting disposal costs. The capability for recovery and reuse or residue treatment is highly dependent upon the particular site.

## CASE STUDY #2: REFINERY SLOP OIL SAMPLES

Slop oils containing hazardous compounds such as benzene, toluene, and chromium, were treated in the Carver-Greenfield pilot plant in 1985. The slop oil was separated into solid, water, and oil-soluble (indigenous oil) products. The original slop oil was "unsuitable" for use in landfills, and the final solids product “met prevailing requirements” for use in nonhazardous landfills (US EPA 1992a).

Approximately 18 kg (40 lb) of slop oil sample (containing 12% solids) was slurried with 36 kg (80 lb) of solvent containing ≈0.2 kg (0.4 lb) of a surfactant (US EPA 1992a). The slurry was subjected to evaporation to reduce water content to less than 1%, then cooled and centrifuged.

The centrifuge cake was split into two samples. Solvent was extracted once from sample 1 and twice from sample 2. Following extraction, the samples were deoiled in a vacuum oven at 149°C (300°F). See table C.5 for a summary of the product's composition following deoiling. The liquid portion (solvent/indigenous oil mixture) was treated by evaporation and steam stripping, resulting in an indigenous oil product with less than 0.8% solvent (US EPA 1992a).

Two points should be noted regarding this study. First, the ratio of the carrier solvent to solids was significantly high. Specifically, 36 kg (80 lb) of solvent was used, although the initial slop oil sample contained only ≈2.3 kg (5 lb) of solids. This is higher than the 2.3 to 4.5 kg (5 to 10 lb) of solvent per 0.45 kg (1 lb) ratio of the waste solids guideline for the SITE demonstration (US EPA 1992a). Second, no information is presented permitting quantification of amounts removed through evaporation and

**Table C.5**

Carver - Greenfield: Refinery Slop Oil Sample Composition by Weight Percent

| | | | | Decontaminated Solids | |
|---|---|---|---|---|---|
| Component | Slop Oil | Fresh Slurry | Centrifuge Cake | Treated Once | Treated Twice |
| Solids | 12.0 | 3.9 | 48 | 97.8 | 99.6 |
| Indigenous Hydrocarbons | 16.0 | 5.3 | 4.6 | 2.0 | 0.2 |
| Water | 72.0 | 23.7 | <1 | <0.1 | <0.1 |
| Solvent | 0.0 | 67.1 | 47 | <0.1 | <0.1 |

Compiled from US EPA 1992a, D-2 and D-3

deoiling effects, as opposed to amounts removed through solvent extraction. This causes difficulties in evaluating the chemical/solvent extraction step. In evaluating the Carver-Greenfield Process, however, the total removal effected by all of the process steps should be considered.

## CASE STUDY #3: PETROLEUM SLUDGE

Dissolved air flotation sludge, American Petroleum Institute (API) separator bottoms, tank bottoms, biosludge, and primary/secondary emulsions from a refinery were mixed in different proportions to produce three feeds for treatability studies in 1984 (see table C.6). The feed was subjected to two simulated stages of evaporation.

During the water evaporation, approximately 12% by weight of the indigenous oil was vaporized and condensed with the water.

The dry slurry was centrifuged. The centrifuged solids were then reslurried in a carrier solvent (boiling point 188° to 193°C (370° to 380°F))

**Table C.6**
Carver - Greenfield: Petroleum Sludge

| Parameter | Feed Mix A | Feed Mix B | Feed Mix C |
|---|---|---|---|
| Feed Materials | (%) | (%) | (%) |
| DAF Sludge | 26.5 | 26.5 | 26.5 |
| API Separator Bottoms | 0.0 | 0.0 | 2.9 |
| Tank Bottoms | 0.0 | 12.3 | 0.0 |
| Bio Sludge | 10.2 | 10.2 | 11.3 |
| Emulsions | 63.3 | 51.0 | 56.4 |
| Feed Components | (%) | (%) | (%) |
| Solids | 1.5 | 1.6 | 2.1 |
| Water | 44.0 | 35.2 | 44.7 |
| Indigenous oil | 57.6 | 63.8 | 53.2 |
| Total Feed Components | 103.1 | 100.6 | 100.0 |
| Deoiled Solids Product | (%) | (%) | (%) |
| Solids | 95.5 | 96.0 | 94.4 |
| Water | <0.1 | <0.1 | <0.1 |
| Indigenous oil | 4.3 | 3.8 | 5.4 |
| Solvent | <0.1 | <0.1 | <0.1 |

Compiled from US EPA 1992a, D-5 and D-7. Feed Mix B 100.6% total feed components subtotal was corrected from 99.6% cited in original source.

and filtered using a Buchner funnel. The filtered solids were deoiled in a 121°C (250°F) vacuum oven. The final solids product contained 3.8 to 5.4% indigenous oil, reduced from initial indigenous oil-to-solid ratios in the range of 25 to 40 (US EPA 1992a). Hazardous compounds present in the feed such as benzene and phenol were reportedly removed from the solids (US EPA 1992a).

According to Dehydro-Tech Corporation (DTC), the light indigenous oil distilled during drying was used as the carrier oil/solvent (Trowbridge 1992).

It should be noted that this bench-scale test does not exactly simulate the Carver-Greenfield Process as used in other studies (US EPA 1992a), since the initial stages essentially involved subjecting the waste to evaporation before slurrying with the carrier solvent. Again, no information was presented regarding amounts removed through evaporation (in the first stages) or deoiling (in a heated, vacuum oven), as opposed to amounts removed through the effects of extraction.

## CASE STUDY #4: AMTRAK PETROLEUM HYDROCARBON AND PCB CONTAMINATION

A former diesel fuel storage area was found to be contaminated with petroleum hydrocarbons — soil and separate-phase contamination. As reported by the developer, PCBs were found in concentrations of 5 to 360 ppm in the separate-phase hydrocarbons (Trowbridge, Holcombe, and Kollitides 1991). Three samples containing 20 to 28% oil, 67 to 74% solids, and 4 to 6% water were treated through triple extraction/dehydration in laboratory-scale testing. Resulting solids, in all cases, contained <1% oil, and water was undetectable (Trowbridge, Holcombe, and Kollitides 1991). In one sample, a PCB (Aroclor 1260) concentration was reduced from 7 ppm to less than 0.11 ppb in treated solids (Trowbridge, Holcombe, and Kollitides 1991).

## CASE STUDY #5: DRYING APPLICATIONS

Several case studies of the Carver-Greenfield System's use in commercial operations are summarized here.

*Wool Scouring Wastes.* Wool scouring wastes and industrial activated sludge were treated in a pilot-scale plant to assess treatment capabilities, as well as to evaluate recovery of lanolin from the oil in the wool scouring waste. An 18:1 ratio of wool scouring waste-to-activated sludge was treated (see table C.7). According to the developer, solvent is not detected in the effluent water condensate (US EPA 1992a).

**Table C.7**
Carver - Greenfield: Commercial Plant

| Parameter | Activated Sludge | | Wool Scouring Waste | |
|---|---|---|---|---|
| Feed Composition Design | (wt%) | | (wt%) | |
| Water | 85.0 | | 94.4 | |
| Solids | 12.8 | | 3.4[1] | |
| Oil | 2.2 | | 2.2[1] | |
| Feed Flow Rate Design | (kg/h) | (lb/h) | (kg/h) | (lb/h) |
| Total Flow | 983 | 2,213 | 17,699 | 39,822 |
| Typical Operation Solids Product | | (%) | | |
| Water | | 0.6 | | |
| Solvent | | 0.9 | | |

1. Expected feed composition was changed to 1.5% solids and <1% oil (versus design values of 3.4 and 2.2, respectively), and equipment was modified to accommodate this change.

US EPA 1992a, Appendix D

The plant is designed with five evaporation stages. In the first three stages, water is evaporated from the wool scouring waste without any solvent present. Solvent is added in the fourth and fifth stages, and the activated sludge is added in the fifth stage. The stages were designed to operate at 49°, 60°, 71°, 82°, and 127°C (120°, 140°, 160°, 180°, and 260°F). Following centrifuging, solids are deoiled in three stages, and lanolin is recovered in a solvent stripping unit.

*Sewage Sludge Drying.* Another application is the drying of sewage sludge. As described by the developer, sewage sludge has a typical energy content of 15,100 kJ/kg (6,500 Btu/lb) of dry solids, but moisture content affects the accessibility of the energy. The Carver-Greenfield Process has been used to dry sewage sludges, thereby producing a fuel with a net available heat of almost 9,300 kJ/kg (4,000 Btu/lb) of solids. The dried sludge is also combustible at high flame temperatures (1,100° to 1,650°C (2,000° to 3,000°F)), which destroy contaminants. Plants have been installed in several cities, including Tokyo and Los Angeles (Holcombe and Kollitides 1991).

In Los Angeles, for example, the Carver-Greenfield Process was a part of the Los Angeles Hyperion Energy Recovery System (HERS) project (Bress, Greenfield, and Haug). The HERS, which converts sludge to gaseous and solid sludge derived fuels (Bress, Greenfield, and Haug), was completed in 1987 (Harrison and Crosse 1991). The project was implemented to eliminate disposal of the sludge at ocean outfalls.

After three years of operation, the HERS system enabled the city to move from ocean disposal to 100% landfill and then to 100% beneficial reuse of the sludge (Harrison and Crosse 1991). The reuse and disposal options are intended to be applied flexibly and vary month to month (Harrison, Smith, and Crosse 1990). For example, sludge reuse and disposal allocations in July, 1989 were 30.6% land application, 31.7% energy recovery, 19.9% landfilling, and 17.8% landfill cover. Those in August, 1989 were 46.8% land application, 22.2% energy recovery, with 14.5% used in landfills, and 16.5% landfill cover (Harrison, Smith, and Crosse 1990). By January of 1991, energy recovery had increased to 56% of total sludge, while 28% was applied to land, 9% was chemically fixed, and 7% was used as compost (Harrison and Crosse 1991).

The HERS system produces gaseous fuel (methane) from anaerobic digestion. As the system is designed, the digested sludge (containing 5% solids) would be centrifuged to produce a cake consisting of 18 to 20% solids, which would be ground and fluidized with a light solvent oil for treatment by the Carver-Greenfield Process (Bress, Greenfield, and Haug). According to the developer, three Carver-Greenfield Process trains were designed to treat 360 dry tonne/day (400 dry ton/day) of sewage sludge from a population of 3.5 million (Holcombe and Kollitides 1991; Trowbridge 1992). Another source, however, indicated that the design capacity of the Carver-Greenfield facility was 241 dry tonne/day (265 dry ton/day) and that this capacity had not yet been achieved. (According to DTC, the design basis was 120 dry tonne/day (135 dry ton/day) per treatment train with one on standby (Trowbridge 1992)). As a result, indirect rotary steam dryers were recommended in order to augment the Carver-

Greenfield drying capacity (Harrison, Smith, and Crosse 1990). The ongoing work on the Carver-Greenfield Process is reportedly aimed at achieving a final system capacity of 136 to 182 dry tonne/day (150 to 200 dry ton/day) (Harrison, Smith, and Crosse 1990).

Four multieffect evaporation stages were designed to operate under varied temperatures and pressures (under vacuum). Approximately 790 kJ are expended per kilogram of water evaporated (340 Btu/lb), compared to 3,020 kJ/kg (1,300 Btu/lb) for indirect steam heated sludge dryers and 4,690 kJ/kg (2,020 Btu/lb) for direct-fired sludge dryers. The resulting solids, which have a heating value of around 13,000 kJ/kg (5,600 Btu/lb), are then sent to a fluidized bed gasification system to generate power. Because pathogens are destroyed, the potential for illnesses is reduced in comparison with that of other processes that convert sludge to compost that have been reported (Bress, Greenfield, and Haug).

The capital and operating costs reported for the entire HERS system (including the Carver-Greenfield Process) in 1984 were ≈$187/dry tonne ($170/dry ton) of sludge fed. Of this, $24/dry tonne ($22/dry ton) were operating costs, which were low because of the revenue from the generated electricity (Bress, Greenfield, and Haug).

*Peat Drying.* Another application is drying peat, which contains bitumen, a commercial by-product. Peat is usually sun-dried, a method dependent upon good weather. Laboratory- and pilot-scale studies disclosed that iso-octanol was found to provide the best balance between bitumen extraction, fluidization, and costs (Holcombe and Kollitides 1991).

*Alum Sludge Drying.* The Carver-Greenfield Process is being used to dry alum sludge under an agreement between Foster Wheeler USA and the Contra Costa Water District (CCWD). The water district needed a method of disposing of alum sludge, while Foster Wheeler was planning to construct a cogeneration plant to generate electricity and steam for an oil refinery (Lau 1991).

A two-stage system was installed. Low-pressure steam is used to dehydrate the alum sludge. The pressure reduction allows vaporization to occur at lower-than-normal temperatures. Temperatures increase from 60°C (140°F) in the first stage to 124°C (255°F) in the last, while the vacuum is decreased. Because of the energy efficiency of the Carver-Greenfield Process, total energy use can be less than 700 kJ/kg (300 Btu/lb) of water evaporated, compared to 4,650 kJ/kg (2,000 Btu/lb) for conventional systems (Lau 1991).

Following evaporation, a centrifuge removes half of the oil from the oil-solids slurry. The sludge cake is then sent to a deoiler, which reduces carrier oil to 0.2 to 0.5% on the solids. During the first year of operations, the extracted water had to be treated by the plant sewage system because of carrier oil contamination. Modifications were then made in an attempt to allow the water to be used in the plant cooling system instead (Lau 1991).

Unlike the sludge treated by the Carver-Greenfield unit that was installed at the Los Angeles Hyperion plant, the CCWD sludge does not contain sufficient organic material for burning as fuel. The alum sludge powder was classified as nonhazardous. According to one source, some heavy metals appear to pass through the system, while others (cadmium, beryllium, silver, selenium, and antimony) are either volatilized with the recovered water or filtered by the carbon adsorption vent (Lau 1991). Mass balance studies, however, would be required to characterize the removal process.

During the first 100 days after start up, the plant was operational for only 5 days. Problems included internal gumming, process line plugging, centrifuge seal leaks, centrifuge feed tube failure, and dust emissions from the vent collector. Most of the problems were remedied. Apparently, gumming occurred when the sludge slurry feed fell below 1% solids. As a result, the solids slurry was concentrated to 1.5 to 2% solids. About 570 L (150 gal) of surfactant was added to prevent gumming. The optimal feed solids concentration is 3 to 4%; however, sludge pumping is difficult at concentrations exceeding 3%. To avoid recurrence of start-up problems, the plant is left in continuous operation, sometimes with minimum feed rates of 57 L/min (15 gal/min). When the process is shut down, transfer pipes become clogged and must be physically cleared (Lau 1991).

When the process is operating, 2% alum sludge is dehydrated to 98%+ solids. The thermal breakdown of aluminum hydroxide into aluminum oxides further reduces the weight by 25%. The dried sludge was originally disposed in landfills, costing $220/dry tonne ($200/dry ton) for hauling and filling because the powder required sealed or pneumatic trucks. In addition, the powder had to be treated with water and dust control additives in order to be disposed of in landfills. An agreement was later reached enabling reuse of the sludge powder in concrete and aggregate production, and disposal costs dropped to $83/dry tonne ($75/dry ton) (Lau 1991).

# D

# LOW-ENERGY EXTRACTION PROCESS (LEEP®)

The following case studies present results of bench-scale treatability tests of the LEEP treating a variety of contaminated materials. The data in this summary were provided by the developer (Steiner and Rugg 1992). As observed in Subsection 3.1.2.2, a pilot-scale study is underway; however, no results were available for release at the time of this writing.

The treatability study results were based on leaching tests of soils conducted to gauge the potential effectiveness of full-scale treatment.

## CASE STUDY #1: MANUFACTURED GAS PLANTS (TAR CONTAMINATED SOIL)

During research and development of LEEP, the effectiveness of a variety of solvent and solvent mixtures was tested. The developer currently uses only acetone and a proprietary solvent in the LEEP system. However, the results of one of the case studies which tested other solvent mixtures is presented here to highlight some of the potential issues involved in applying solvent extraction in general.

In the research study, solvent treatment of tar and PAH-contaminated soils was tested. The tested solvents included acetone, acetone/methylene chloride mixture, n-butylamine, acetone/n-butylamine mixture, acetone/diethylamine mixture, solvent "X," solvent "Y," and solvent "Z" (Steiner and Rugg 1992).

A baseline contamination profile was established by averaging the results of gas chromatography/mass spectrometry (GC/MS) analyses (Method 8270) of three sample aliquots. Each aliquot consisted of leachate combined from two soil samples. Based on contaminant solubility in the eight different solvents, six multistage leaching schemes were designed. Each experiment consisted of 10 stages. Table D.1 outlines the leaching schemes.

The results are summarized in table D.2 (on page D.3). The highest removal efficiencies were achieved by Scheme 4 (acetone and an unidentified cosolvent mixture) and Scheme 6 (acetone and diethylamine mixture). According to the developer, the elevated tentatively identified compound (TIC) levels resulting from Scheme 5 (acetone and n-butylamine mixture) appeared to be due to some kind of reaction between n-butylamine and the soil (Rugg 1992). Although the targeted compounds were significantly reduced in all cases (>90%), the research indicated that elevated TICs may be an unwanted side effect. In addition, concentrations of some compounds (such as naphthalene, di-n-butylphthalate, and di-n-octylphthalate) were higher in a few end products than they were initially. As with other soil treatment technologies, some of these differences may be attributed to the natural variations in nonhomogenous soil. The research data indicates that site-specific matrix effects should be tested. As noted earlier, the developer of LEEP currently utilizes only acetone and a proprietary solvent (not one of those tested in this case study) in the LEEP process.

## CASE STUDY #2: CLAY SUBSOIL (POLYCHLORINATED BIPHENYL (PCB)- CONTAMINATED SOLIDS).

A soil sample containing a PCB (Aroclor 1260) concentration of 1500 ppm was treated. The soil sample was composed of (by weight) 18% clay

**Table D.1**
LEEP Leaching Schemes

| Stage | Scheme 1 A | Scheme 2 A-Y-A | Scheme 3 A-X-A | Scheme 4 A-Z-A | Scheme 5 A/B | Scheme 6 A/D |
|---|---|---|---|---|---|---|
| 1 | Acetone | Acetone | Acetone | Acetone | Ac/N-butyl | Ac/diethyl |
| 2 | Acetone | Acetone | Acetone | Acetone | Ac/N-butyl | Ac/diethyl |
| 3 | Acetone | Y | X | Z | Ac/N-butyl | Ac/diethyl |
| 4 | Acetone | Y | X | Z | Ac/N-butyl | Ac/diethyl |
| 5-10 | Acetone | Acetone | Acetone | Acetone | Ac/N-butyl | Ac/diethyl |

**Table D.2**
LEEP Residual Soil Analyses After 10th Stage of Leaching With Various Solvents

| Semivolatile Compounds | Baseline | Concentration (μg/g) Scheme 1 A | Scheme 2 A-Y-A | Scheme 3 A-X-A | Scheme 4 A-Z-A | Scheme 5 A/B | Scheme 6 A/D |
|---|---|---|---|---|---|---|---|
| Naphthalene | 10.20 | 0.75 | 0.30 | 0.27 | 0.20 | 10.78 | 0.07 |
| Acenaphthylene | 8.27 | 0.29 | 0.06 | | 0.13 | 0.37 | |
| Acenaphthene | 1.73 | | | | | | |
| Fluorene | 6.02 | 0.07 | | | | | |
| Phenanthrene | 19.20 | 0.37 | 0.14 | 0.13 | 0.15 | 1.33 | |
| Anthracene | 7.17 | 0.14 | | | 0.07 | | |
| Di-N-butylphthalate | 2.46 | 4.00 | 2.00 | 3.51 | 1.75 | 1.05 | 0.29 |
| Fluoranthene | 11.03 | 0.12 | 0.05 | | 0.08 | | |
| Pyrene | 22.06 | 0.20 | 0.08 | 0.07 | 0.09 | | |
| Benzo(a)anthracene | 9.76 | 0.07 | | | | | |
| Bis(2-ethylhexyl)phthalate | 9.08 | 0.16 | 1.05 | 2.69 | 1.06 | 0.14 | 0.14 |
| Chrysene | 12.25 | 0.09 | | | 0.08 | | |
| Di-N-octylphthalate | 0.73 | | | 1.18 | | | |
| Benzo(b)fluoranthene | 11.50 | 0.05 | | | | | |
| Benzo(k)fluoranthene | 4.46 | 0.05 | | | | | |
| Benzo(a)pyrene | 9.96 | | | | | | |
| Ideno(1,2,3-cd)pyrene | 2.73 | | | | | | |
| Di-benzo(a,h)anthracene | 0.43 | | | | | | |
| Benzo(g,h,i)perylene | 2.46 | | | | | | |
| Subtotal BNs[1] | 151.50 | 5.91 | 2.61 | 6.84 | 2.20 | 12.77 | 0.23 |
| TICs[2] (Other than BNs) | 230.61 | 34.89 | 43.32 | | 0.21 | 2261.58 | |
| Total | 382.11 | 40.80 | 45.93 | 6.84 | 2.41 | 2274.35 | 0.23 |
| BN Removal (%) | - | 96.1% | 98.3% | 95.5% | 98.5% | 91.6% | 99.8% |
| TIC Removal (%) | - | 84.9% | 81.2% | - | 99.9% | neg. | - |
| Total Removal (%) | - | 89.3% | 88.0% | 98.2% | 99.4% | neg. | 99.9% |

1. BNs are the family of targeted chemical compounds classified as base neutral soluble semivolatiles and listed above.
2. TICs are Tentatively Identified Compounds which were not specifically targeted (reported as sum of all TICs in the sample).

neg. = negative removal efficiency (treated concentration greater than initial concentration)
NOTE: Leaching solvents for each scheme were as follows: acetone (scheme 1), acetone and unidentified co-solvent mixtures (schemes 2, 3, & 4), acetone and n-butylamine (scheme 5), and acetone and diethylamine (scheme 6).

(particles, <5 μm), 52% silt (particles from 5 μm to 50 μm), and 30% sand (particles from 50 μm to 500 μm). See table D.3 for the resulting leaching removal efficiencies.

## CASE STUDY #3: INDUSTRIAL LANDFILL (PCB-CONTAMINATED SOLIDS)

A sample of landfill material contaminated with PCBs (Aroclor 1254 and 1260), semivolatile organics, and heavy metals was treated. The sample was composed of (by weight) 40% water and soil [17% clay (<5 μm), 55% silt (5 to 50 μm), and 28% larger particles (>50 μm)]. The initial concentrations of contaminants on a dry weight basis were as follows: 10,584 ppm PCBs and 1,078 ppm semivolatiles [the semivolatiles concentrations were 469 ppm TICs and 609 ppm specifically identified base neutral (BN) soluble semivolatiles].

The developer calculated the total amount of contaminants removed as the sum of the amounts of contaminants removed from each stage. Specifically, the "concentration remaining in the soil" after each stage was calcu-

**Table D.3**

LEEP Multistage PCB Leaching Results From Clay

| Stage # | PCB Removed Per Stage (lg) | Total PCB Removed (lg) | Stage Leaching Efficiency[1] (%) | Overall Leaching Efficiency[2] (%) |
|---|---|---|---|---|
| 1 | 19,155 | 19,155 | 78 | 78 |
| 2 | 3,678 | 22,833 | 67 | 93 |
| 3 | 1,318 | 24,150 | 73 | 98 |
| 4 | 245 | 24,395 | 51 | 99 |
| 5 | 157 | 24,552 | 67 | 99.7 |
| 6 | 69 | 24,621 | 88 | 99.9 |
| 7,8,9 | 10 | 24,631 | 99+ | 99.9+ |

1. The Stage Leaching Efficiency is the amount of contaminant removed in each stage divided by the initial amount of contaminant in that stage.
2. The Overall Leaching Efficiency is the total (cumulative) amount leached divided by the initial amount of contaminant in the sample.

lated based on the dry sample weight, assuming that all of the contaminants (including contaminants in the residual leaching solvent) were those in the soil. Actual soil concentrations, based on separate analyses, were presented only for the 12th stage, where the concentrations were determined by analyzing the air-dried residual soil (Steiner and Rugg 1992). No information regarding mass balance closure was presented, and therefore, an independent review of the data was not possible (for example, to quantify potential contaminant remnants in the residual solvent or to serve as a cross-check on the calculated staged removal efficiencies).

Table D.4 presents the efficiencies effected in removing BN, and table D.5 (on page D.6), those in removing PCBs. After three stages of treat-

**Table D.4**

LEEP Base Neutral (Semivolatile) Leaching: Industrial Landfill

| Stage # | BN Removed Per Stage (μg) | Total BN Removed (μg) | BN conc. Remaining in Soil (ppm)[1] | Stage Leaching Efficiency (%)[2] | Overal Leaching Efficiency (%)[3] |
|---|---|---|---|---|---|
| 1 | 2,218 | 2,218 | 387 | 36 | 36 |
| 2 | 1,178 | 3,396 | 269 | 30 | 56 |
| 3 | 520 | 3,916 | 217 | 19 | 64 |
| 4 | 232 | 4,148 | 194 | 11 | 68 |
| 5 | 112 | 4,260 | 183 | 6 | 70 |
| 6 | 52 | 4,312 | 177 | 3 | 71 |
| 7 to 12 | 224 | 4,536 | 15[4] | 12 | 98 |

| | |
|---|---|
| Initial BN concentration in the sample (excluding TICs) | 609 ppm |
| Wet sample weight | 16.7 g |
| Dry sample weight | 10 g |
| Leaching solvent volume | 40 mL |
| Stage leaching time | 1 hr |

1. The BN Concentrations shown in this column are calculated based on the dry sample weight, assuming all BNs, including those in the residual leaching solvent, to be associated with the solids.
2. The Stage Leaching Efficiency is the amount of contaminant removed in each stage divided by the initial amount of contaminant in that stage.
3. The Overall Leaching Efficiency is the total (cumulative) amount leached divided by the initial amount of contaminant in the sample.
4. The BN concentration in the sample after the 12th stage was determined by analyzing the air dried residual soil.

**Table D.5**
LEEP PCBs Leaching: Industrial Landfill

| Stage # | PCBs Removed Per Stage (lg) | Total PCBs Removed (lg) | PCBs conc. Remaining in Soil (ppm)[1] | Stage Leaching Efficiency (%)[2] | Overal Leaching Efficiency (%)[3] |
|---|---|---|---|---|---|
| 1 | 71,500 | 71,500 | 4593 | 60.86 | 60.86 |
| 2 | 30,300 | 101,800 | 1568 | 65.89 | 86.60 |
| 3 | 10,850 | 112,650 | 483 | 69.19 | 95.89 |
| 4 | 3,640 | 116,290 | 119 | 75.33 | 98.98 |
| 5 | 635 | 116,925 | 56 | 53.27 | 99.52 |
| 6 | 232 | 117,157 | 33 | 41.56 | 99.72 |
| 7 to 12 | 277 | 117,434 | 4.8[4] | 85.23 | 99.96 |

| | |
|---|---|
| Initial BN concentration in the sample (excluding TICs) | 10,584 ppm |
| Wet sample weight | 16.7 g |
| Dry sample weight | 10 g |
| Leaching solvent volume | 40 mL |
| Stage leaching time | 1 hr |

1. The PCBs Concentrations shown in this column are calculated based on the dry sample weight, assuming all PCBs, including those in the residual leaching solvent, to be associated with the solids.
2. The Stage Leaching Efficiency is the amount of contaminant removed in each stage divided by the initial amount of contaminant in that stage.
3. The Overall Leaching Efficiency is the total (cumulative) amount leached divided by the initial amount of contaminant in the sample.
4. The PCBs Concentration in the sample after the 12th stage was determined by analyzing the air dried residual soil.



ment, the overall removal of BN was only 64% (up from 36% in the first stage), while the overall removal of PCBs increased from 61% in the first stage to 96% in the third stage. After the 12th stage, when the soil was analyzed, removal of 98% BN was achieved, and 99.96% of PCBs (Steiner and Rugg 1992; Mallach 1992).

## CASE STUDY #4: REFINERY SLUDGES (OIL & GREASE, SEMIVOLATILE ORGANIC CONTAMINATION, VOLATILE ORGANIC CONTAMINATION, AND METALS — AS, CD, CU, PB, ZN, NI, V, AND CR)

Two refinery sludge samples were treated. Sludge #1 consisted of the solids fraction from a rainwater impoundment. Its initial composition was 33% solids, 60% water, and 7% oil and grease. Sludge #2 consisted of an emulsion filter cake from slop recovery waste having an initial composition of 47% solids, 28% water, and 25% oil & grease. The sludges were contaminated with various semivolatile compounds, including anthracene, benzo-a-anthracene, benzo-a-pyrene, bis-2-ethylhexyl-phthalate, chrysene, naphthalene, phenanthrene, and pyrene, along with volatile organics and metals (Steiner and Rugg 1992).

Table D.6 presents the efficiencies effected in removing semivolatiles, and table D.7 (on page D.8), those in removing oil and grease. No data regarding removal of the volatile compounds and metals were presented.

**Table D.6**

LEEP Semivolatile Organic Compounds Leaching: Refinery Sludges

| | Sludge #1 | | Sludge #2 | |
|---|---|---|---|---|
| Stage # | Stage Efficiency[1] | Overall Efficiency[2] | Stage Efficiency[1] | Overall Efficiency[2] |
| 1 | 71.1 | 71.1 | 61.9 | 61.9 |
| 2 | 75.3 | 92.7 | 43.1 | 78.0 |
| 3 | 59.7 | 97.2 | 74.4 | 94.4 |
| 4 | 40.5 | 98.4 | 51.7 | 97.8 |
| 5 | 29.9 | 98.9 | 32.9 | 98.1 |
| 6 | | 99.9+ | 57.4 | 99.2 |
| 7 | | | 45.7 | 99.5 |
| 8 | | | | 99.9+ |

1. The Stage Efficiency is the amount of contaminant removed in each stage divided by the initial amount of contaminant in that stage.
2. The Overall Efficiency is the total (cumulative) amount leached divided by the initial amount of contaminant in the sample.

**Table D.7**
LEEP Oil And Grease Leaching: Refinery Sludges

| | Sludge #1 | | Sludge #2 | |
|---|---|---|---|---|
| Stage # | Stage Efficiency[1] | Overall Efficiency[2] | Stage Efficiency[1] | Overall Efficiency[2] |
| 1 | 26.6 | 26.6 | 23.9 | 23.9 |
| 2 | 57.1 | 68.4 | 42.9 | 58.4 |
| 3 | 62.9 | 88.4 | 50.7 | 78.5 |
| 4 | 65.8 | 96.0 | 48.8 | 88.9 |
| 5 | 66.7 | 98.4 | 51.5 | 94.7 |
| 6 | | 99.9+ | 64.1 | 98.2 |
| 7 | | | 99.9+ | |

1. The Stage Efficiency is the amount of contaminant removed in each stage divided by the initial amount of contaminant in that stage.
2. The Overall Efficiency is the total (cumulative) amount leached divided by the initial amount of contaminant in the sample.



The developer noted, however, that materials were decontaminated below regulatory limits (Steiner and Rugg 1992).

The most significant gains in the efficiency in semivolatile removal were again observed in the first three stages of treatment. Efficiencies increased from the 62 to 71% range in the first stage to the 94 to 97% range after three stages. Efficiencies exceeding 99.9% were achieved in treating Sludge #1 after 6 stages, while 8 stages were required to attain this level in treating Sludge #2.

Efficiencies in removing the oil and grease increased more gradually initially than those in removing semivolatiles. For example, efficiencies in removing oil and grease of only 24 to 27% were attained after one stage, while those in removing semivolatiles were 62 to 71%. But 99.9% efficiency was attained in removing the oil and grease from Sludge #1 after 6 stages, and from Sludge #2, after 7 stages.

## CASE STUDY #5: WAUKEGAN HARBOR SEDIMENT (PCB CONTAMINATION)

During the early development of the LEEP system a treatability study on sediment from Waukegan Harbor demonstrated 99.9% removal of a PCB. The PCB-contaminated (Aroclor 1242) sample had an initial concentration of 33,641 ppm PCBs (dry weight basis) and consisted of 42% water by weight (Steiner and Rugg 1992).

# E
# NUKEM DEVELOPMENT (NKD) PROCESSES

This appendix presents the results of NKD Process pilot plant treatability studies carried out on (1) seven soils, all contaminated with polychlorinated biphenyls (PCBs), (2) two refinery feedstocks, (3) an American Petroleum Institute (API) separator sludge, and (4) a slop oil sludge.

The following soils were tested:

- Four soils taken at varying depths from the General Electric Rose Superfund Site in Pittsfield, Massachusetts (RS-10, RS-14, RS-16, RS-18);
- One soil from a Superfund site in Texas;
- One soil from a Superfund site in New Jersey; and
- One soil from a site in Mississippi.

These samples were selected to enable study of a wide range of key properties that could be expected to affect the performance of the process.

Table E.1 (on page E.2) presents the properties of the seven soils before treatment. Table E.2 (on page E.2) indicates the particle size distributions of three of the soils taken from the Rose Superfund Site.

Extraction tests were performed in bench-scale equipment utilizing 30-gram samples of contaminated soil. The contaminated soil, as received, was placed in a container and mixed with a small amount of reagent and a measured amount of fresh solvent. Solvent-to-soil loadings varied from 1:1 to 2:1 by weight. The mixture was agitated for approximately three minutes and allowed to settle. The solvent was then separated from the soil by decantation, and a sample of the soil was drawn for PCB analysis. This proce-

**Table E.1**
NKD Process Properties of Test Soils

| Test Soil | Appearance | Physical Properties (As Received) | | | |
|---|---|---|---|---|---|
| | | Moisture (Wt %) | Oil & Grease (Wt %) | Dry Specific Gravity | Initial PCBs (mg/kg) |
| Massachusetts Soil | | | | | |
| • RS-10 | Sandy | 8.4 | NA | NA | 3,300 |
| • RS-14 | Clayey | 13.5 | 0.3 | 1.61 | 771 |
| • RS-16 | Sandy | 11.5 | 0.4 | 2.50 | 1,147 |
| • RS-18 | Clayey | 26.5 | 3.6 | 1.32 | 3,130 |
| Mississippi Soil | Clayey | 15.0 | NA | NA | 41 |
| New Jersey Soil | Silty | 15.0 | NA | NA | 1,944 |
| Texas Soil | Clayey, high plasticity | 25.0 | Trace | NA | 50 |

**Table E.2**
NKD Process Particle Size Distribution of Selected Test Soils

| Particle Size (mm) | Rose Site Test Soils (Wt %) | | |
|---|---|---|---|
| | RS-14 | RS-16 | RS-18 |
| >2 | 18.8 | 49.25 | 17.9 |
| 2.00 -0.861 | 7.4 | 13.50 | 19.9 |
| 0.841-0.420 | 4.9 | 7.10 | 12.6 |
| 0.420-0.250 | 3.9 | 4.10 | 7.6 |
| 0.250-0.177 | 2.8 | 2.20 | 3.6 |
| 0.177-0.149 | 1.8 | 1.10 | 1.8 |
| 0.149-0.125 | 1.8 | 1.00 | 2.2 |
| 0.105-0.074 | 5.9 | 2.20 | 3.5 |
| <0.074 | 49.7 | 18.10 | 28.9 |

dure, constituting a single stage of extraction, was repeated as many times as necessary to achieve a desired overall level of decontamination.

Stage-by-stage batch extraction results are set forth in table E.3 (on page E.3). Although not all tests were continued to completion, each of the test soils was determined to be treatable to residual PCBs levels of less than 2 mg/kg. With the exception of PCBs and oil and grease concentrations, the properties of each soil were essentially unaffected by the treatment process. In particular, moisture and clay plasticity were essentially unchanged.

The concentrations of PCBs in the Rose Superfund Site samples ranged from 771 to 3,131 mg/kg. After four stages of extraction, the residual PCBs concentrations in these samples were all less than 25 mg/kg. As the results for Sample RS-10 illustrate, when processing was carried to completion, residual PCBs levels were reduced to less than 2 mg/kg. Essentially all oil and grease were removed from each of the Rose Superfund Site samples.

**Table E.3**
Batch Extraction Performance of NKD Soil Washing Process
(PCB Concentration in mg/kg)

| Extraction Stage | Massachusetts Soils | | | | Mississippi Soil | New Jersey Soil | Texas Soil |
|---|---|---|---|---|---|---|---|
| | RS-10 | RS-14 | RS-16 | RS-18 | | | |
| 0 | 3,300 | 771 | 1,147 | 3,131 | 41 | 1,944 | 50 |
| 1 | 330 | 245 | 368 | 298 | 11 | 291 | 10 |
| 2 | 40 | 67 | 163 | 58 | 4.5 | 48 | 0 |
| 3 | 4 | 23 | 56 | 25 | 1.6 | 8 | - |
| 4 | 0 | 8 | 22 | 7 | - | 1 | - |

After two stages of extraction, no residual PCBs could be detected in the sample of the Texas soil, although this soil was particularly difficult to treat (clay material with a high moisture content and a high plasticity).

The pilot countercurrent extraction column, described in Subsection 3.13.3, was used to treat two refinery waste streams, an API separator sludge, and a slop oil sludge. The pilot plant also contained a complete fractional distillation column for the processing, recovery, and recycle of solvent, along with a number of dewatering devices, including a plate and frame filter press, a belt filter press, and a vacuum filter.

The basic characteristics of the two refinery feedstocks are listed in table E.4.

The API separator sludge was taken directly from the API separator pit of a major Gulf Coast refinery. The sample of slop sludge was taken from a waste oil dumping pit at another major Gulf Coast refinery.

**Table E.4**
Characteristics of Two Refinery Feedstocks

| Component | API Separator (Primary) | Slop Sludge (Secondary) |
|---|---|---|
| % Water | 81% | 78% |
| % Solids | 9% | 16% |
| % Oil | 10% | 6% |

Characteristics of the product oils recovered from the samples of API separator and slop oil sludge, as determined from the form of true boiling point curves, varied significantly. The oil recovered from the sample of slop sludge was heavier than that recovered from that of API separator sludge.

Residual hydrocarbon characteristics of the solids and the water produced in the treatment of API separator sludge are presented in table E.5. Included for comparison purposes are applicable solid and water best demonstrated available control technology (BDAT) standards. Only four compounds are present at or above the levels of analytical detection, and only one of these compounds, phenol, is regulated. The BDAT requirements were met. Basic chromatographic data suggested that similar results could be expected for the products of slop sludge processing. Table E.5 shows that the NKD Process effects a high level of removal of oil from both water and solids present in raw API separator sludge.

## Table E.5
Fate of Organics in Pilot-Scale Treatment of API Separator Sludge

| | Raw Sludge | Filtered Solids | Filtered Water | BDAT Solid | Standard Water |
|---|---|---|---|---|---|
| Benzene | 434 | ND | ND | 14 | 0.011 |
| Ethyl Benzene | 142 | ND | ND | 14 | 0.011 |
| Toluene | 359 | ND | ND | 14 | 0.011 |
| Xylene | 939 | ND | ND | 22 | 0.011 |
| Phenol | 16.41[1] | 3.35[1] | 0.58[1] | 3.6 | 0.047 |
| bis-(2-Chloroethyl) Ether | 1.16 | ND | ND | - | - |
| 4-Methylphenol (p-cresol) | 1.08 | ND | 0.09 | - | - |
| 2-Nitrophenol | 5.03 | ND | ND | - | - |
| 2,4- Dimethylphenol | 5.55 | ND | ND | - | - |
| bis-(2-Chloroethoxy) Methane | 1.5 | ND | ND | - | - |
| 2,4-Dichlorophenol | 10.40 | ND | ND | - | - |
| Naphthalene | 1.00 | ND | ND | 42 | 0.033 |
| 4-Chloroaniline | 7.78 | ND | ND | - | - |
| 2-Methylnaphthalene | 79.5 | 2.72 | 0.3 | - | - |
| 4-Chloro-3-methylphenol | 1.7 | ND | ND | - | - |
| 2,4,5-Trichlorophenol | 2.39 | ND | ND | - | - |
| 2-Chloronaphthalene | 1.78 | ND | ND | - | - |
| 2-Nitroaniline | 20.8 | 0.79 | 0.09 | - | - |
| 3-Nitroaniline | 22.2 | ND | 0.08 | - | - |
| 2,4 Dinitrophenol | 8.02 | ND | ND | - | - |
| Dibenzofuran | 1.11 | ND | ND | - | - |
| 2,4-Dinitrotoluene | 20.3 | ND | ND | - | - |
| 4,-Nitrophenol | 43.3 | 1.62 | ND | - | - |
| Fluorene | ND | ND | ND | - | 0.054 |
| -Chlorophenyl-4 -Nitroaniline | 1.06 | ND | ND | - | - |
| 4,6-Dintro-2-methylphenol | 102.5 | ND | ND | - | - |
| N-Nitrosodiphenylamine | 5.76 | ND | ND | - | - |
| Phenanthrene | ND | ND | ND | 34 | 0.039 |
| Anthrancene | 16.7 | ND | ND | 28 | 0.039 |
| Dibutyl Phthalate | ND | ND | 0.10 | 3.6 | 0.06 |
| Pyrene | 2.9 | ND | ND | 14 | 0.011 |
| Benzo(a)anthracene | ND | ND | ND | 20 | 0.043 |
| Chrysene | 2.91 | ND | ND | 1 | 50.043 |
| 3,3'-Dichlorobenzidine | 1.37 | ND | ND | - | - |
| bis-(2-ethylhexyl) Phthalate | 5.31 | ND | ND | 7.3 | 0.043 |
| di-n-octyl Phthlate | ND | ND | 0.08 | - | - |
| Benzo(a)pyrene | ND | ND | ND | 12 | 0.047 |
| Indeno(1,2,3-cd)pyrene | 1.88 | ND | ND | - | - |
| Benzo(g,h,i)perylene | 3.10 | ND | ND | - | - |

1. The phenol readings reported are highly suspect. Extremely low levels of phenol were reported in the raw sludge (ppm). Phenol is highly soluble in water; it is also soluble in NKD process solvent. In spite of this and the very large quantities of both present, the analytical data suggest limited dissolution of phenol.

Characteristics of metals of the solids produced in the treatment of API separator sludge in the NKD Process pilot plant are presented in table E.6. These were preliminary results, but showed that stabilization of solids at and below BDAT standards is readily achievable.

**Table E.6**

Leachability of Metals in Samples of Pilot Scale Processed API Separator Solids

| | | TCLP, ppm | |
|---|---|---|---|
| | Metal Content of Dry Solids, ppm | Stabilized [1,2] Sample | Detection Limit |
| Ag | < 1 | < 0.01 | 0.01 |
| As | 5.2 | 0.01 | 0.002 |
| Ba | 940 | 1.6 | 0.02 |
| Cd | 5.7 | < 0.01 | 0.01 |
| Cr | 220 | < 0.02 | 0.02 |
| Hg | 35 | < 0.001 | 0.001 |
| Ni | 75 | < 0.02 | 0.02 |
| Pb | 100 | < 0.025 | 0.025 |
| Se | 1.0 | < 0.005 | 0.005 |

1. Test sample was stabilized with 0.5 units of stabilizing chemicals per unit of filter wet solids. Results clearly indicate that substantially less stabilizing chemical is needed to comply with the BDAT requirements. Optimization work is underway now to determine minimum quantity of stabilizing chemicals required for BDAT level stabilization.

2. Stabilization resulted in a 15 percent increase in the volume of the solids.